Lower Secondary
Mathematics
7

Ric Pimentel
Frankie Pimentel
Terry Wall

The Publishers would like to thank the following for permission to reproduce copyright material.

Photo acknowledgements

p.1 *t* © Agefotostock/Alamy Stock Photo, *b* © Oleg Seleznev/stock.adobe.com; **p.77** *t* © Ntguilty/Alamy Stock Vector, *b* © Jakezc/stock.adobe.com; **p.165** © Sergii Figurnyi/stock.adobe.com; **p.171** *t* © Agefotostock/Alamy Stock Photo, *b* © Donyanedomam/stock.adobe.com; **p.192** © nuttakit – Fotolia; **p.196** Paula/stock.adobe.com

Acknowledgements

Cambridge International copyright material in this publication is reproduced under licence and remains the intellectual property of Cambridge Assessment International Education.

End of section review questions (and sample answers) have been written by the authors. In assessments, the way marks are awarded may be different.

Third-party websites and resources referred to in this publication have not been endorsed by Cambridge Assessment International Education.

Every effort has been made to trace all copyright holders, but if any have been inadvertently overlooked, the Publishers will be pleased to make the necessary arrangements at the first opportunity.

Although every effort has been made to ensure that website addresses are correct at time of going to press, Hodder Education cannot be held responsible for the content of any website mentioned in this book. It is sometimes possible to find a relocated web page by typing in the address of the home page for a website in the URL window of your browser.

Hachette UK's policy is to use papers that are natural, renewable and recyclable products and made from wood grown in well-managed forests and other controlled sources. The logging and manufacturing processes are expected to conform to the environmental regulations of the country of origin.

Orders: please contact Hachette UK Distribution, Hely Hutchinson Centre, Milton Road, Didcot, Oxfordshire, OX11 7HH. Telephone: +44 (0)1235 827827. Email education@hachette.co.uk Lines are open from 9 a.m. to 5 p.m., Monday to Friday. You can also order through our website: www.hoddereducation.com

ISBN: 978 1 3983 0194 8

© Ric Pimentel, Frankie Pimentel and Terry Wall 2021

First published in 2004
Second edition published in 2011
This edition published in 2021 by
Hodder Education,
An Hachette UK Company
Carmelite House
50 Victoria Embankment
London EC4Y 0DZ

www.hoddereducation.co.uk

The authorised representative in the EEA is Hachette Ireland, 8 Castlecourt Centre, Dublin 15, D15 XTP3, Ireland (email: info@hbgi.ie)

Impression number 10 9 8

Year   2025

All rights reserved. Apart from any use permitted under UK copyright law, no part of this publication may be reproduced or transmitted in any form or by any means, electronic or mechanical, including photocopying and recording, or held within any information storage and retrieval system, without permission in writing from the publisher or under licence from the Copyright Licensing Agency Limited. Further details of such licences (for reprographic reproduction) may be obtained from the Copyright Licensing Agency Limited, www.cla.co.uk

Cover photo © radachynskyi - stock.adobe.com

Illustrations by Integra Software Services Pvt. Ltd, Pondicherry, India

Typeset in Palatino LT Std Light 11/13 by Integra Software Services Pvt. Ltd, Pondicherry, India

Printed in Great Britain by Bell and Bain Ltd, Glasgow

A catalogue record for this title is available from the British Library.

Contents

The units in this book have been arranged to match the Cambridge Lower Secondary Mathematics curriculum framework. Each unit is colour coded according to the area of the syllabus it covers:

- ■ Number
- ■ Geometry & Measure
- ■ Statistics & Probability
- ■ Algebra

How to use this book — v

Introduction — vii

Section 1

Unit 1	Addition, subtraction, multiplication and division	2
Unit 2	Properties of two-dimensional shapes	8
Unit 3	Data collection and sampling	13
Unit 4	Area of a triangle	19
Unit 5	Order of operations	25
Unit 6	Algebra beginnings – using letters for unknown numbers	28
Unit 7	Organising and presenting data	36
Unit 8	Properties of three-dimensional shapes	54
Unit 9	Multiples and factors	64
Unit 10	Probability and the likelihood of events	70
Section 1 Review		73

Contents

Section 2

Unit 11	Rounding and estimation – calculations with decimals	78
Unit 12	Mode, mean, median and range	87
Unit 13	Transformations of two-dimensional shapes	95
Unit 14	Manipulating algebraic expressions	109
Unit 15	Fractions, decimals and percentages	116
Unit 16	Probability and outcomes	130
Unit 17	Angle properties	135
Unit 18	Algebraic expressions and formulae	153
Unit 19	Probability experiments	158
Unit 20	Introduction to equations and inequalities	162
Section 2 Review		168

Section 3

Unit 21	Sequences	172
Unit 22	Percentages of whole numbers	179
Unit 23	Visualising three-dimensional shapes	185
Unit 24	Introduction to functions	192
Unit 25	Coordinates and two-dimensional shapes	198
Unit 26	Squares, square roots, cubes and cube roots	205
Unit 27	Linear functions	211
Unit 28	Converting units and scale drawings	221
Unit 29	Ratio	230
Unit 30	Graphs and rates of change	237
Section 3 Review		247

Glossary	250
Index	255

How to use this book

To make your study of Cambridge Lower Secondary Mathematics as rewarding as possible, look out for the following features when you are using the book:

History of mathematics

These sections give some historical background to the material in the section.

These aims show you what you will be covering in the unit.

LET'S TALK
Talk with a partner or a small group to decide your answer when you see this box.

Worked example

These show you how you could approach answering a question.

KEY INFORMATION
These give you hints or pointers to solving a problem or understanding a concept.

These highlight ideas and things to think about.

This book contains lots of activities to help you learn. The questions are divided into levels by difficulty. Green are the introductory questions, amber are more challenging and red are questions to really challenge yourself. Some of the questions will also have symbols beside them to help you answer the questions.

Exercise 15.2

1 Work out the answer to the following calculations. Show your working clearly and simplify your answers where possible.

a $\frac{2}{5}+\frac{1}{6}$

b $\frac{7}{12}+\frac{1}{5}$

c $\frac{9}{14}-\frac{2}{7}$

d $\frac{3}{13}-\frac{3}{26}$

e $\frac{1}{8}+\frac{5}{16}-\frac{5}{24}$

f $\frac{13}{18}-\frac{8}{9}+\frac{1}{6}$

2 Sadiq spends $\frac{1}{5}$ of his earnings on his mortgage. He saves $\frac{2}{7}$ of his earnings. What fraction of his earnings is left?

3 The numerators of two fractions are hidden as shown.

$$\frac{\square}{8} + \frac{\square}{5} = \frac{23}{40}$$

The sum of the two fractions is $\frac{23}{40}$. Calculate the value of both numerators.

Look out for these symbols:

🟢 This green star icon shows the thinking and working mathematically (TWM) questions. This is an important approach to mathematical thinking and learning that has been incorporated throughout this book.

Questions involving TWM differ from the more straightforward traditional question-and-answer style of mathematical learning. Their aim is to encourage you to think more deeply about the problem involved, make connections between different areas of mathematics and articulate your thinking.

🖩 This indicates where you will see how to use a calculator to solve a problem.

❌ These questions should be answered without a calculator.

🔗 This tells you that content is related to another subject.

🔊 This tells you that content is available as audio. All audio is available to download for free from www.hoddereducation.com/cambridgeextras

▶ There is a link to digital content at the end of each unit if you are using the Boost eBook.

Introduction

This series of books has been written by experienced teachers who have lived or worked in schools and with teachers from countries around the world, including the United Kingdom, Spain, Germany, France, Turkey, South Africa, Malaysia and the U.S.A.

Students and teachers in these countries come from a variety of cultures and speak many different languages, as well as English. Sometimes cultural and language differences can make understanding difficult.

However, mathematics is almost a universal language. 28 + 37 = 65 will be understood in many countries where English is not the first language. A maths book written in Japanese will include algebra equations with x and y, for example.

We should also all be very aware that much of the mathematics you will learn in this series of books was first discovered, and built upon, by mathematicians from China, India, Arabia, Greece and Mesopotamia (modern day Iraq).

Our present number system originated in India with the Brahmi numerals in about 300 BCE.

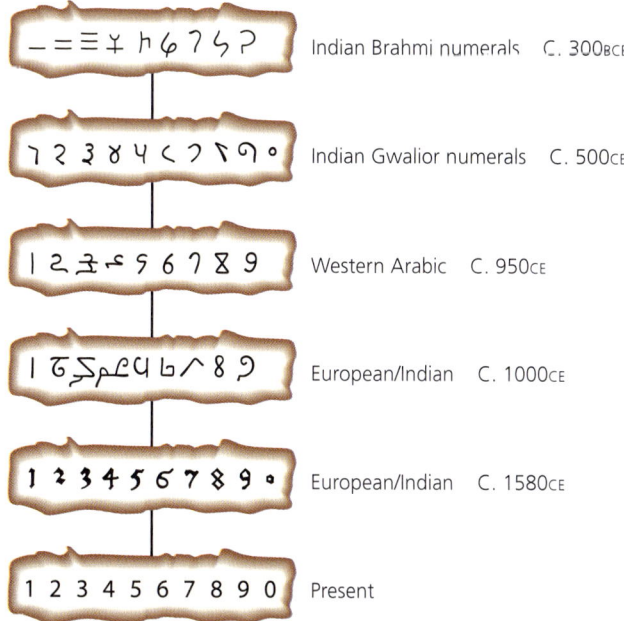

Introduction

Most early mathematics was simply game play and problem solving. It wasn't until later that this maths was applied to building, engineering and sciences of all kinds. Mathematicians study maths because they enjoy it as fun in itself.

We hope that you will enjoy the mathematics you learn in this series of books. Sometimes the ideas will not be easy to understand at first – that should be part of the fun! Ask for help if you need it, but try hard first.

Try to learn by thinking and working mathematically (TWM). This is an important feature of the book. Thinking and working mathematically can be divided into the following characteristics:

Thinking and working characteristic	Definition
Specialising	Choosing an example and checking if it satisfies or does not satisfy specific mathematical criteria.
Generalising	Recognising an underlying pattern by identifying many examples that satisfy the same mathematical criteria.
Conjecturing	Forming mathematical questions or ideas.
Convincing	Presenting evidence to justify or challenge a mathematical idea or solution.
Characterising	Identifying and describing the mathematical properties of an object.
Classifying	Organising objects into groups according to their mathematical properties.
Critiquing	Comparing and evaluating mathematical ideas, representations or solutions to identify advantages and disadvantages.
Improving	Refining mathematical ideas or representations to develop a more elegant approach or solution

Where you see this icon ⭐ it shows you that you will be thinking and working mathematically.

Writing down your thoughts and workings helps to develop your mathematical fluency. By thinking carefully about how you explain your ideas you may, while justifying an answer, be able to make wider generalisations. Discussing different methods with other students will also help you compare and evaluate your mathematical ideas. This will lead to you understand why some methods are more effective than others in given situations. Throughout you should always be forming further mathematical questions and presenting other ideas for thought.

Many students start off by thinking that mathematics is just about answers. Although answers are often important, posing questions is just as important. What is certainly the case is that the more you question and understand, the more you will enjoy mathematics.

Ric Pimentel, Frankie Pimentel and Terry Wall, 2021

SECTION 1

History of mathematics – The development of number

In what is now The Democratic Republic of the Congo in Africa, bones have been discovered with marks cut into them. The Ishango bones, as they are called, are believed to be about 20 000 years old. The marks are thought to be an early form of tally marks. The tally marks may have been used for counting time, such as numbers of days or cycles of the moon, or for keeping records of numbers of animals.

A tallying system has no place value, which makes it hard to show large numbers.

The earliest system like ours dates to about 3100 BCE in Egypt.

Many ancient texts, for example from Babylonia (modern Iraq) and Egypt, used zero. Egyptians used the word *nfr* to show a zero balance in accounting. Indian texts used a Sanskrit word, *shunya*, to refer to the idea of the number zero. By the 4th century BCE, the people of South-Central Mexico began to use a true zero. It was represented by a shell picture and became part of Mayan numerals.

Addition, subtraction, multiplication and division

- Estimate, add and subtract integers, recognising generalisations.
- Estimate, multiply and divide integers including where one integer is negative.

Being able to carry out calculations without the help of a calculator is an important skill. This unit looks at some methods for doing calculations and checking them.

Mental skills

There are of course written methods for adding and subtracting numbers. However mental strategies are useful too.

One method is to look for **number bonds** to 10 or 100.

Worked examples

1. Without a calculator add the following two numbers: **52+68**.

   The units **2** and **8** form a number bond to **10**. Therefore the calculation can be worked out by adding together **50+60+10**=120.

   This can also be visualised using a number line.

   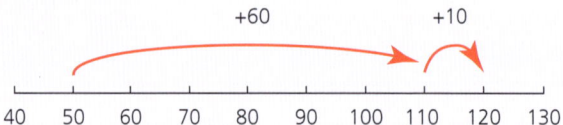

2. Without using a calculator subtract the following numbers: **146−36**.

   The units of both numbers are the same, so **6−6**=0. The calculation can therefore be worked out using the remaining digits, i.e. **140−30**=110.

   On the number line below, 146−36 can be seen to be the same as 140−30.

   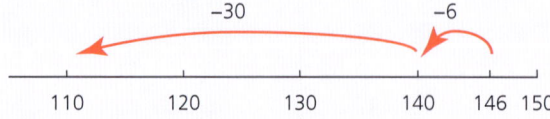

A second method is to use borrowed numbers.

> **Worked examples**
>
> 1. Without a calculator add the following two numbers: $79+32$.
>
>   79 is very nearly 80. Therefore an easier calculation would be to add $80+32=112$.
>
>   But because 1 was added to 79 to make the 80, it must now be subtracted from the answer. Therefore $79+32=111$.
>
> 2. Without using a calculator subtract the following numbers: $118-44$.
>
>   118 can be **rounded** up to 120 as this will make the calculation easier.
>
>   It now becomes $120-44=76$. But as 2 was added to 118 it must be subtracted from the answer. Therefore $118-44=74$.

> **LET'S TALK**
>
> What do you notice happens to the answer as the number being subtracted decreases by 1 each time?

Subtracting negative numbers

Look at the following number pattern:

$10-4=6$
$10-3=7$
$10-2=8$
$10-1=9$
$10-0=10$

If the pattern is continued, the number subtracted becomes negative and the answer continues to increase by 1.

$10-(-1)=11$
$10-(-2)=12$
$10-(-3)=13$

It can be seen that the result of subtracting a negative number is the same as adding it as a positive number, i.e. $10-(-3)=10+3=13$.

Therefore subtracting a negative number makes the answer bigger.

Estimating

When adding or subtracting larger numbers, the two mental methods shown above aren't as easy to use, so it is always good to be able to estimate roughly what the answer should be.

To estimate an answer, round each of the numbers to make the calculation easier.

SECTION 1

> Numbers are rounded to the nearest 1000, 100 or 10 to make the estimation easier.

> Remember to write down your thinking when you work through the exercises. This will help you think and work mathematically.

Worked examples

1. Estimate the answer to the following calculation:

   $286 + 407$

   Rounding 286 to the nearest hundred becomes 300.
   Rounding 407 to the nearest hundred becomes 400.
   Therefore an estimate for $286 + 407$ is $300 + 400 = 700$.

2. A pupil does the following calculation in his head:

   $1121 + 788 - 210 - (-592)$

   and says that the answer is 1291.
   Explain why this answer must be incorrect.
   Rounding each of the numbers gives the following approximation:

   $1000 + 800 - 200 - (-600) = 2200$

   The estimation gives an answer of 2200. As 1291 is just over half this amount it is likely to be incorrect.

Exercise 1.1

Work with a friend and without a calculator. Calculate the answers to the following questions.

1. a $38 + 12$ e $59 + 31$ i $54 - 48$
    b $45 + 25$ f $34 - 12$ j $81 - 55$
    c $66 + 34$ g $45 - 34$
    d $23 + 27$ h $98 - 65$

2. At the start of a journey a car's distance meter reads 1421 km.
   A family set off on a journey that is 289 km long.
   What will the car's distance meter read at the end of the journey?

3. a $123 + 27$ f $43 - (-17)$
    b $334 + 56$ g $176 - (-34)$
    c $67 + 153$ h $87 - (-233)$
    d $42 + 478$ i $-77 - (-57)$
    e $567 + 23$ j $-81 - (-161)$

4. A glass with capacity 230 ml is filled from a jug containing 725 ml of water. How much water is left in the jug after:
   a one glass of water is poured
   b two glasses of water are poured
   c three glasses of water are poured?

1 Addition, subtraction, multiplication and division

5. A 3×3 magic square contains each of the numbers 1–9 arranged in such a way that each row, column and diagonal add up to the same total.
    a Copy the 3×3 grid opposite and enter the numbers 1–9 so that it becomes a magic square.
    b i) Complete a magic square using the numbers from 51 to 59.
       ii) How does the total of each row, column and diagonal compare with the original magic square in part (a)? Explain why this happens.
    c i) Complete a magic square using the numbers from −4 to +4.
       ii) How does the total of each row, column and diagonal compare with the original magic square in part (a)? Explain why this happens.

6. A lift states that the maximum load it can carry is 500 kg.
   Six people want to get in the lift.
   The masses of the first five people are 62 kg, 88 kg, 79 kg, 91 kg and 86 kg.
   The sixth person says, 'I have a mass of 92 kg so it won't be safe for me to go in the lift as well.'
   Convince them it is safe to go in the lift.

Multiplication and division

Remember to write down your thinking when you work through the exercises. This will help you think and work mathematically.

You will already be familiar with multiplication and division of positive **integers**. The following exercise will recap some of this.

Exercise 1.2

Do these questions in your head, without looking at a multiplication grid or using a calculator. Write down your answers.

1. Multiply the following pairs of numbers.
   a 150×5
   b 14×70
   c 16×40
   d 130×50
   e 12×30
   f 180×7
   g 90×16
   h 70×330
   i 24×60
   j 160×8
   k 2700×7
   l 240×90

2. Divide the following pairs of numbers.
   a 124÷4
   b 168÷8
   c 2400÷6
   d 320÷80
   e 4900÷70
   f 560÷8
   g 440÷44
   h 3600÷60
   i 4800÷80

3. A child buys five packets of sweets each containing 80 sweets. How many sweets does he buy in total?

SECTION 1

4. A bar of chocolate is made of square-shaped pieces.
   The bar is four pieces wide and 12 pieces long. Amir needs 336 pieces to make some chocolate cakes. How many bars should he buy?

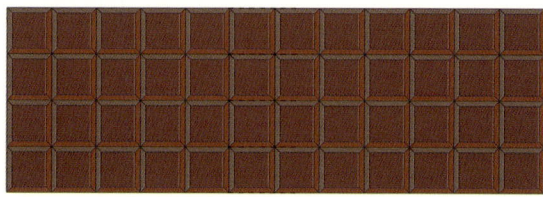

5. A tricycle is like a bicycle, but with three wheels. A shop has a stock of 19 tricycles and 112 bicycles. How many wheels are there in total?

6. For a school concert 360 chairs are arranged in rows of 20.
   How many rows are there?

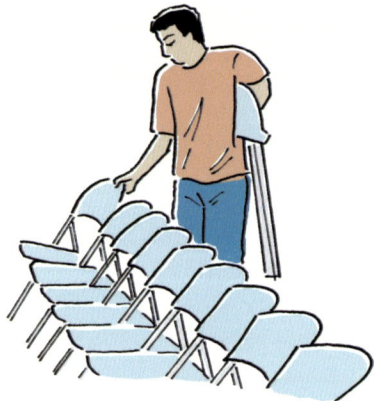

Multiplication and division by negative numbers

Look at the number pattern produced by the multiplications below.

$3 \times 4 = 12$
$2 \times 4 = 8$
$1 \times 4 = 4$
$0 \times 4 = 0$

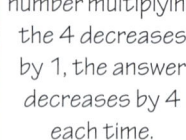

Notice how as the number multiplying the 4 decreases by 1, the answer decreases by 4 each time.

If the multiplication is continued, the answers follow the same pattern.

$(-1) \times 4 = -4$
$(-2) \times 4 = -8$
$(-3) \times 4 = -12$

You can see that if a positive number is multiplied by a negative number the answer is negative too.

As $-3 \times 4 = -12$ it follows that $-12 \div 4 = -3$.

1 Addition, subtraction, multiplication and division

> Remember to write down your thinking when you work through the exercises. This will help you think and work mathematically.

Therefore dividing a negative number by a positive number produces a negative answer.

Multiplying a negative number by a positive number produces a negative answer

Dividing a negative number by a positive number produces a negative answer

Exercise 1.3

1 a Copy and complete the multiplication table below. One is completed for you.

×	12	35	48	125
−2	−24			
−6				
−12				

b Using the table above, rewrite the multiplications to form a division. For example, $-24 \div 12 = -2$

2 Below are six cards, each with a number on them.

−3, 2, 6, 12, −18, −9

Select three cards each time. Use multiplication or division to produce as many correct calculations as possible.

e.g. $\boxed{2} \times \boxed{6} = \boxed{12}$

3 $36 \times 18 = 648$

State the answer to each of the following calculations and explain how the above calculation was used to work it out.

    a $648 \div 18$ c 18×18 e -18×9
    b -36×18 d $-648 \div 9$ f $-1296 \div 36$

LET'S TALK
With a friend discuss what other calculations you could work out using $36 \times 18 = 648$.

Now you have completed Unit 1, you may like to try the Unit 1 online knowledge test if you are using the Boost eBook.

2 Properties of two-dimensional shapes

- Identify, describe and sketch regular polygons, including reference to sides, angles and symmetrical properties.
- Understand that if two 2D shapes are congruent, corresponding sides and angles are equal.
- Know the parts of a circle.

Polygons

Any closed, two-dimensional shape made up of straight lines is called a **polygon**.

Both triangles and quadrilaterals belong to the family of **polygons**. If all the sides are the same length and all the interior angles are equal, the shape is called a **regular polygon**. A square and an equilateral triangle are examples of regular polygons.

The names of some other common polygons are:

5 sides **pent**agon   6 sides **hex**agon   8 sides **oct**agon

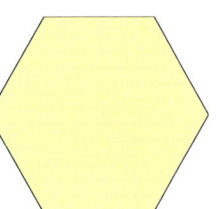
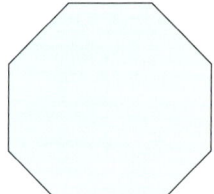

> **LET'S TALK**
>
> The prefixes 'pent', 'hex' and 'oct' are often used to describe the numerical property 5, 6 and 8 respectively. Discuss with your partner examples of when this occurs in everyday language.
>
> Are there examples where these rules do not apply?

Exercise 2.1

 1 Copy and complete the following table to **classify** the properties of quadrilaterals. The first line has been started for you.

	Rectangle	Square	Parallelogram	Kite	Rhombus	Trapezium
Opposite sides equal in length	Yes					
All sides equal in length						

8

2 Properties of two-dimensional shapes

	Rectangle	Square	Parallelogram	Kite	Rhombus	Trapezium
All angles right angles						
Both pairs of opposite sides parallel						
Diagonals equal in length						
Diagonals intersect at right angles						
All angles equal						

2. The following diagram shows two **congruent** isosceles right-angled triangles. The two triangles can be arranged side by side to form a larger isosceles right-angled triangle as shown:

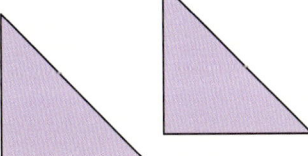 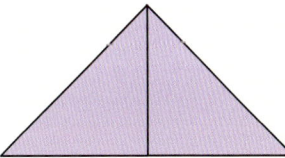

Using only two congruent isosceles right-angled triangles, sketch which quadrilaterals can be formed by placing the triangles together so their sides are fully in contact.

3. The diagram below shows two equilateral triangles.

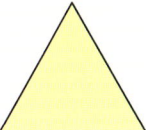 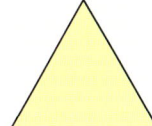

Which quadrilateral(s) can be formed by arranging the two triangles so that their sides are fully in contact? Use a sketch to justify your answer.

Regular polygons have a number of interesting properties to do with line symmetry and **rotational symmetry**.

SECTION 1

Worked example

Draw an equilateral triangle, mark on its lines of symmetry and state its order of rotational symmetry.

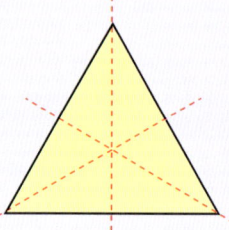

The equilateral triangle has a rotational symmetry of order 3.

Exercise 2.2

1. a Copy and complete the following table to **classify** regular polygons and their symmetry properties.

Name of regular polygon	Number of sides	Shape	Number of lines of symmetry	Order of rotational symmetry
	3	△		
Square				
	5	⬠		
	6			
Octagon				
		⬡		

b   What do you notice about the relationship between the number of sides a regular polygon has, the number of lines of symmetry and its order of rotational symmetry?

2 The geometric pattern below shows a square and equilateral triangle inside a circle. All their **vertices** are on the **circumference** of the circle. The square and equilateral triangle have a common point P.

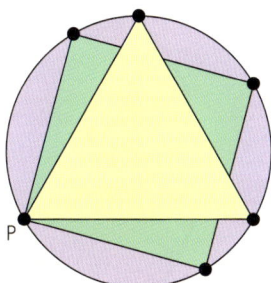

a   How many lines of symmetry does the pattern have?
b   What is the order of rotational symmetry of the pattern?

3 The geometric pattern below shows a regular hexagon and equilateral triangle inside a circle. All their vertices are on the circumference of the circle. The regular hexagon and equilateral triangle have the common points X, Y and Z.

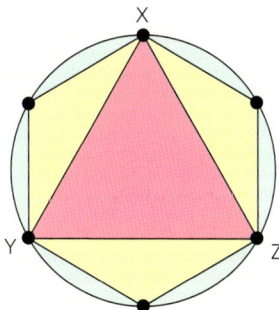

a   How many lines of symmetry does the pattern have?
b   What is the order of rotational symmetry of the pattern?

4 A designer wants to design a new company logo. She decides that the logo will consist of a circle with two regular polygons drawn inside it. The vertices of the regular polygons all lie on the circumference of the circle.
a   If the final logo has four lines of symmetry and a rotational symmetry of order four, which possible regular polygons has the designer used?
b   Is there another possible combination of two regular polygons, other than those you stated in (a) above, that the designer could have used? Give a **convincing** answer.
c   Sketch a possible logo drawn by the designer.

The circle

There are special words to describe different parts of a circle. The diagram below shows some of the main parts of the circle and gives their names.

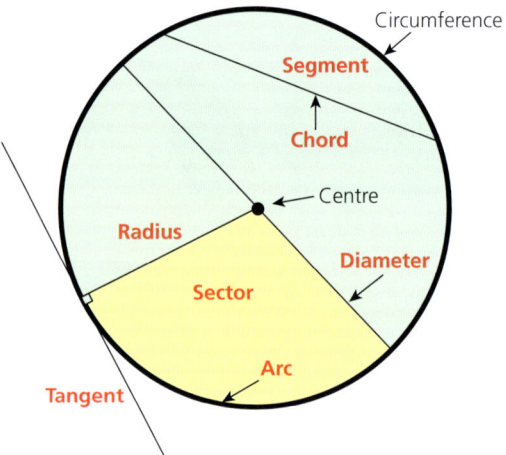

Exercise 2.3

1. Copy and complete these sentences to give the definitions of the terms shown in the diagram above.
   a. A line that is always the same distance from a single fixed point is called a _____.
   b. The perimeter of a circle is called its _____.
   c. A straight line from the centre of a circle to the circumference of the circle is called a _____. The plural of the word is _____.
   d. A straight line across a circle which starts and ends at two points on the circumference is called a _____.
   e. A chord which passes through the centre of a circle is called a _____.
   f. A line which forms part of the circumference of a circle is called an _____.
   g. The area enclosed by two radii and an arc is called a _____.
   h. The area enclosed by an arc and a chord is called a _____.
   i. A straight line which touches the circumference of a circle is called a _____.
   j. The angle between a radius and a tangent to a point on the circumference is _____.

2. Write a sentence linking the two words given below. The first one has been done for you.
   a. Radius and diameter *The radius is half of the diameter length.*
   b. Diameter and chord
   c. Arc and radii
   d. Sector and segment

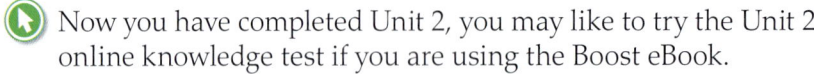 Now you have completed Unit 2, you may like to try the Unit 2 online knowledge test if you are using the Boost eBook.

3 Data collection and sampling

- Select and trial data collection and sampling methods to investigate predictions for a set of related statistical questions, considering what data to collect.
- Understand the effect of sample size on data collection and analysis.

 ## Types of data

Pieces of information are often called **data**. **Quantitative data** – that is, data that can be measured – falls into two categories: **discrete data** and **continuous data**.

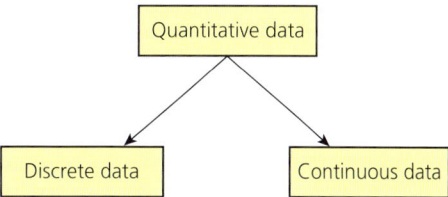

It is important to understand the difference between the two types, as the type of **data** collected often affects the types of graphs that can be drawn to represent it and the calculations that can be carried out.

Discrete data can only take specific values. For example, a survey to find the numbers of students in different classes can only produce values that are whole numbers. 30 students in a class would be a possible data value but 30.2 would not.

Not all data is restricted to specific values. For example, if data is collected about a sprinter's 100 m running time, a result of 10.2 seconds may be collected. But this result is only accurate to one **decimal place**. A more precise stopwatch may give the time as 10.23 seconds, but this in itself is only accurate to two decimal places.

Even more accurate timings may give the result as:

10.228 seconds	(to three decimal places)
10.2284 seconds	(to four decimal places)
10.228 39 seconds	(to five decimal places)

and so on.

> **LET'S TALK**
> Can you and a partner think of at least one example of each type of data?

SECTION 1

Data that can take any value (usually within a range) is known as **continuous data**. The timing of an event is an example of continuous data.

Not all **data** though can be measured numerically. This type of data is often known as categorical data. **Categorical data** is put into groups (or categories). Examples include hair colour, ethnicity and football team supported.

Exercise 3.1

1. Decide which of the following types of data are discrete, which are continuous and which are categorical.
   a. The mass of a baby (kg).
   b. The type of transport students use to get to school.
   c. The number of people attending a concert.
   d. The shoe size of students in your class.
   e. The temperature at midday (°C).
   f. The gender of people at a party.
   g. The type of animal treated by a vet.
   h. The speed of cars passing the school (km/h).
   i. The wingspan of a species of butterfly (mm).
   j. The volume of water in a drinking bottle (cm³).
   k. The number of pages in a book.
   l. The height of students in a class (cm).

Collecting data

Data is often collected in response to a problem.

For example:
- Your school canteen may wish to know what sort of food to sell.
- An athletics club may want to know which athletes perform consistently well.
- Your teacher may want to find out which topics to focus on for revision.

> The term **population** here does not mean everybody in the country. Depending on what the data collection is about, population could mean everybody in your school or even just everybody in your class.

In all these cases, for the answers to be of any use and valid, the method of data collection must be planned properly. This will include deciding how many people are going to be included in the survey. Often not everybody (the **population**) needs to be asked. Asking the whole population can be too time consuming or too expensive, so usually a **representative sample** of the population is asked. The number of people in the population asked is known as the **sample size**.

3 Data collection and sampling

Sampling

If data is not going to be collected from a whole population, then the **sample** taken from it must be **representative** of the population, i.e. the aim is that the results from the sample should be similar to the results of the whole population.

Exercise 3.2

Using an alphabetical list of all the students in your class, try the following methods of selecting a representative sample.

1. a Choose the first 10 students on the list.
   b Has this method picked a representative sample of your class? Justify your answer.
   c If your answer to (b) above was 'no' explain whether the method could have produced a representative sample.
   d If your answer to (b) above was 'yes' explain whether the method might not have produced a representative sample.

2. a Select every
   i) 10th person
   ii) 5th person
   iii) 3rd person
   b Comment on which method(s) produced the most representative sample of your class and which produced the least representative sample.

3. a Number each student 1, 2, 3 etc. then use the random number generator on your calculator to pick some students at random.
   b Comment on whether you think this is a fair way of picking a representative sample. Justify your answer.

> **LET'S TALK**
>
> If a population consists of 1000 people, how many do you think need to be selected in a random sample? Why?

Exercise 3.3

1. A teacher wants to find out what school clubs the students would like the school to offer. The school has a population of 1500 students so he decides that he will ask a sample of the student population. He decides to ask all the students in his next class.
   a Give two **convincing** reasons why his method of sampling may not be representative.
   b Suggest an **improved** way of sampling the student population.

2. A teacher wants to interview a sample of the students in the whole school about what they find difficult in mathematics. She surveys a sample of her own students, by putting their names in a hat and selecting them at random.
   a Explain why the teacher's method will not produce a representative sample of the students in the whole school.
   b Suggest how she can **improve** on her method to get a representative sample.

SECTION 1

3 An athletics club has the following number of members

	Male	Female
Under 14	20	60
Over 14	5	15

They decide on surveying 20 members. As there are four categories, they decide to randomly choose five members from each group.
  a Give two possible reasons why choosing five members from each group will not be a representative sample of the members of the club.
  b Suggest an **improved** way of choosing a representative sample of 20 members of the club.

Data collection

One method of collecting data is to **interview** people. The interviewer asks questions and writes down the answers given.

Another method is to use **questionnaires**. A questionnaire is a printed list of questions which people answer, usually on their own without any help.

Exercise 3.4

1 A school canteen wishes to know what sort of food to sell to students. It carries out interviews with a number of students to find out their opinions. Here is the transcript of the interviewer's conversation with one of the students.
Interviewer: Hello. The school is trying to find out what sort of food it should sell to students. Can you tell me the sorts of foods you like and whether you would consider eating school food?
Student: Hi. At the moment I don't eat school food very often. I usually bring in my own packed lunch although sometimes, if I forget my packed lunch or if my friends are all going to the canteen, then I will eat school food. When I do buy school food, I try not to spend too much, usually only about $3 or $4, but if I've had sport in the morning then I'll be really hungry and probably spend more. I like most foods, like pasta, pizza and curry, but wouldn't want to eat it every day.
  a What problems will the interviewer have when analysing this answer?
  b What is wrong with the question asked by the interviewer?
  c If you were the interviewer, what would your first question have been?
  d The interview lasts 20 minutes. Assuming all the questions are answered properly, give one of the main advantages of this method of data collection.
  e The interviewer interviews 20 students in total. Give one of the main disadvantages of this method of data collection.

3 Data collection and sampling

2 The school in question 1 above also gives out the following questionnaire to the students.

Please tick the appropriate box for each question.
- What year are you in?
  Year 7 ☐ Year 9 ☐ Year 11 ☐
  Year 8 ☐ Year 10 ☐
- What gender are you?
  Male ☐ Female ☐
- On average, how many times a week do you eat in the school canteen?
  Never ☐ Twice ☐ 4 times ☐
  Once ☐ 3 times ☐ Every day ☐
- If you have eaten in the canteen, how would you rate the food?
  Excellent ☐ Good ☐ Satisfactory ☐
  Poor ☐
- What food would you like to see on sale in the canteen? (Tick more than one box if needed.)
  Pizza ☐ Curry ☐ Salad ☐
  Pasta ☐ Fish ☐ Fruit ☐
  Chicken ☐ Vegetarian ☐ Other ☐
- How much would you be prepared to pay for a meal in the canteen?
  Less than $2 ☐ Between $2 and $4 ☐ More than $4 ☐

a Give two benefits of using the questionnaire above. Justify your choices.
b Give two drawbacks of using the questionnaire. Justify your choices.
c If you were designing the questionnaire, what additional questions would you ask?
d The school has 1500 students. The questionnaire is given out to 20 students. Give one criticism of this. Justify your answer.
e What might be a good sample size for this questionnaire? Justify your answer.

You will have seen that both methods of data collection have advantages and disadvantages. Which method is chosen depends on several factors. The following table highlights the main advantages and disadvantages of the two methods.

SECTION 1

	Interview	Questionnaire
Advantages	Detailed answers can be given Interviewer can clarify any misunderstandings Interviews produce a higher response rate Personal Interviewer can pick up on body language or different tone of voice	Can get the opinions of a lot of people, i.e. sample size is likely to be larger Relatively cheap to carry out Relatively quick to fill in Easy to analyse responses Format is familiar to most people Can be used for sensitive topics People have time to think about their answers
Disadvantages	Needs a skilled interviewer so that questions are not biased and so that respondent is relaxed Time consuming and therefore expensive to do Cannot be used for large numbers of people: sample size is likely to be smaller	Not suitable for complex questions Impersonal Some people don't bother returning the questionnaire Those who return the questionnaire may be interested in the topic, so the responses may be biased Questions may be misunderstood Certain questions can be ignored

> **LET'S TALK**
> With a friend discuss and write down what you think is meant by some of these advantages and disadvantages. Give examples if possible, to clarify your thoughts.

 Now you have completed Unit 3, you may like to try the Unit 3 online knowledge test if you are using the Boost eBook.

4 Area of a triangle

- Derive and know the formula for the area of a triangle. Use the formula to calculate the area of triangles and compound shapes made from rectangles and triangles.

Triangles

One of the angles of this triangle is a right angle; this is a right-angled triangle.

Drawing a rectangle around the triangle will help us to work out the formula for the area of a triangle.

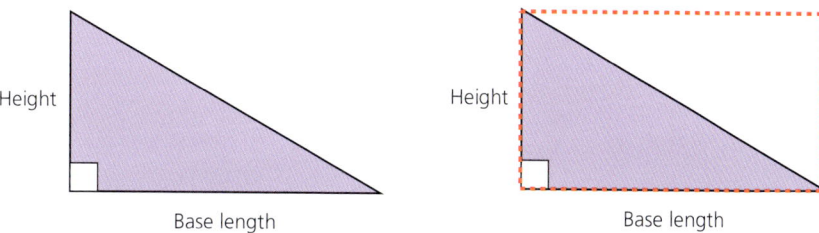

Area of rectangle = base length × height

The area of the triangle is half the area of the rectangle.

Area of triangle = $\frac{1}{2}$ × base length × height

This also works for other triangles.

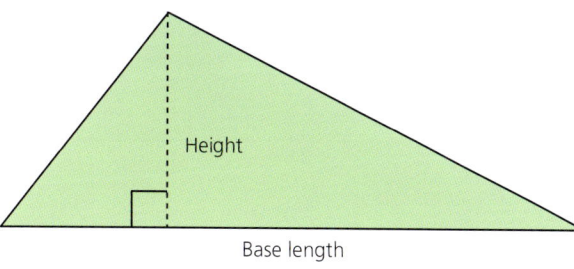

The height is always measured at right angles to the base. This is sometimes known as the **perpendicular height**.

When we draw a rectangle around the triangle, we can see that the area of the triangle is still half that of the rectangle.

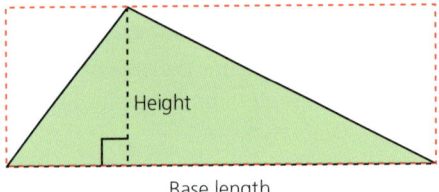

Area of a triangle = $\frac{1}{2}$ × base length × perpendicular height

There will be times when the base of the triangle is not conveniently given as a horizontal side. However, we can use any side of the triangle as the base. We must make sure, though, that we measure the height at right angles to the side chosen as the base. These diagrams demonstrate this.

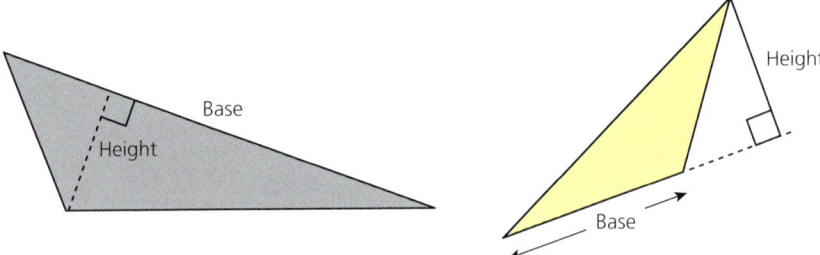

Worked example

Calculate the area of the triangle below.

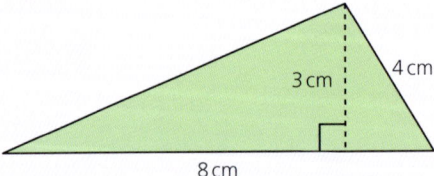

The base length = 8 cm

The perpendicular height = 3 cm

The side length of 4 cm is not needed to calculate the area of the triangle.

Area = $\frac{1}{2}$ × 8 × 3

= 12 cm²

4 Area of a triangle

Exercise 4.1

1 Calculate the area of each of the triangles below:

a 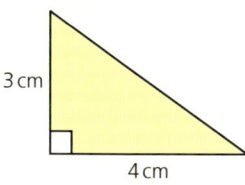 b 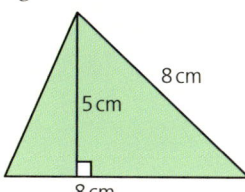 c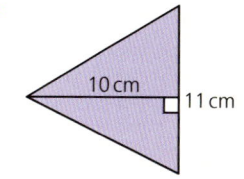

2 Use the formula for the area of a triangle to work out the missing values in the table below:

	Base length	Perpendicular height	Area
a	7.2 cm	4.8 cm	
b	20 cm		100 cm²
c		15 cm	15 cm²
d		11 cm	55 cm²

3 Three triangles P, Q and R share a common base and lie between two parallel lines.
Which of the three triangles (if any) has the biggest area? Give a reason for your answer.

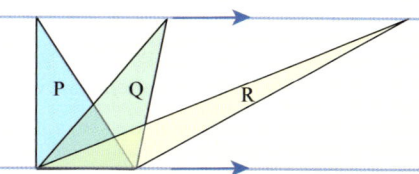

4 a Draw a triangle similar to the one drawn opposite.
   i) Measure the length of side AB and let this be the base of the triangle.
   ii) Draw on your diagram a line which represents the height of the triangle.
   iii) Measure the height of your triangle.
   iv) Calculate the area of your triangle.
  b Draw another triangle identical to the one you drew in part (a).
   i) Measure the length of side BC and let this be the base of the triangle.
   ii) Draw on your diagram, a line which represents the height of the triangle.
   iii) Measure the height of your triangle.
   iv) Calculate the area of your triangle.
  c Draw another triangle identical to the ones you drew in parts (a) and (b).
   i) Measure the length of side AC and let this be the base of the triangle.
   ii) Draw on your diagram, a line which represents the height of the triangle.
   iii) Measure the height of your triangle.
   iv) Calculate the area of your triangle.
  d What do you notice about the area of each of the three triangles you drew?
    Comment carefully on the reasons for any differences or similarities in your answers.

SECTION 1

 5 The line PQ in the grid below is the base of a triangle. Each square in the grid is 1 cm².

  a Copy the diagram and mark on your grid a point R, so that triangle PQR has an area of 8 cm².

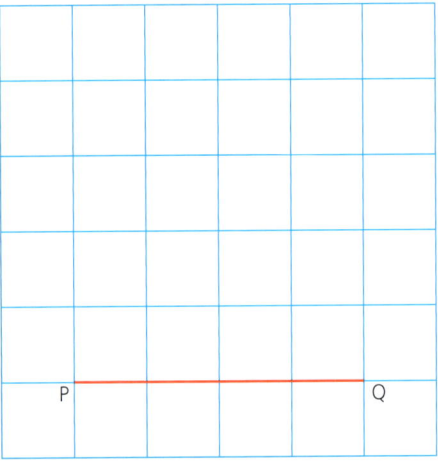

  b Mark on your grid two other possible positions for point R.
  c Comment on the positions of the three possible positions for R that you have chosen.
  d Can you form a **generalisation** from your answers to (b) and (c)?

 6 The diagram below shows a sequence of Sierpinski triangles, which are made of black and white equilateral triangles.

If the first large triangle has an area of 64 cm², calculate the total area of black triangles in each of the other patterns. Show your working clearly.

Compound shapes

A compound shape is one which is made from two or more simpler shapes.

To work out the area of a compound shape it is often easier to work out the area of the simpler shapes first.

4 Area of a triangle

> **Worked example**
>
> Calculate the area of the compound shape below:
>
>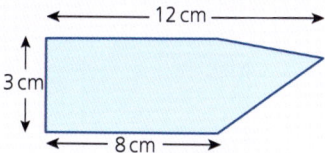
>
> The shape is an example of a compound shape as, although it is a pentagon, it can be split into a rectangle A and a triangle B as shown.
>
>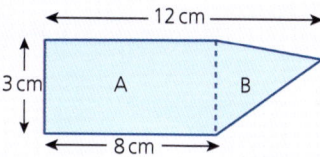
>
> Area of rectangle A: $3 \times 8 = 24 \, cm^2$
>
> Area of triangle B: The 'height' of the triangle can be deduced as it is the difference between the 12 cm and 8 cm measurements, i.e. 4 cm
>
> The area is therefore $\frac{1}{2} \times 3 \times 4 = 6 \, cm^2$
>
> Total area of compound shape is $24 + 6 = 30 \, cm^2$

Exercise 4.2

Calculate the area of each of the shapes below:

 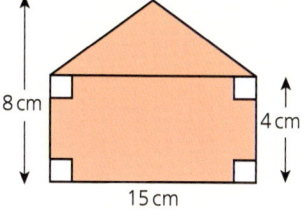

 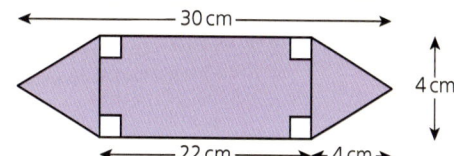

 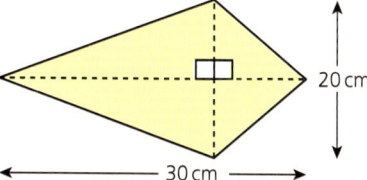

SECTION 1

 4 Two congruent (identical) right-angled triangles are placed inside a square as shown. The square has a side length of 8 cm.
If half the area of the square is not covered by the triangles, calculate the length x of each triangle. Show all your reasoning clearly.

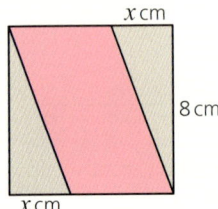

 5 Four congruent right-angled triangles are arranged to form two squares as shown:

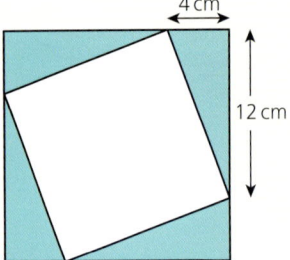

a What is the area of the smaller square as a fraction of the larger square?
b Assuming the size of the larger square stays the same, what could the dimensions of each of the four congruent triangles be so that the area of the smaller square is half that of the larger square?
c Is it possible for the area of the smaller square to be less than half that of the larger square? Give a **convincing** reason for your answer.

> Now you have completed Unit 4, you may like to try the Unit 4 online knowledge test if you are using the Boost eBook.

5 Order of operations

- Understand that brackets, positive indices and operations follow a particular order.

Order of operations

The order in which mathematical calculations are done depends on the operations being used.

Look at this calculation:

$6 + 3 \times 2 - 1$

Carrying out the calculation from left to right would give an answer of 17.

However, if you do the calculation on a calculator, the answer it gives is 11.

This is because mathematical operations are carried out in a particular order:

- **B**rackets   Any operation in brackets is done first.
- **I**ndices   A number raised to a power (**index**) is done next.
- **D**ivision and/or **M**ultiplication   Multiplications and divisions are done next. Their order does not matter.
- **A**ddition and/or **S**ubtraction   Additions and subtractions are carried out last. Again, their order is not important.

A way of remembering this order is with the shorthand **BIDMAS**.

The correct answer to the calculation $6 + 3 \times 2 - 1$ is 11, because the 3×2 must be done first, followed by addition of the 6 and subtraction of the 1.

The 1 can be subtracted before the 6 is added; the answer is still 11.

LET'S TALK
Why is it necessary to have an agreed order for calculations?

Worked examples

1 Calculate $7 + 4 \times 9 - 8$.

$7 + 4 \times 9 - 8$
$= 7 + 36 - 8$ *The multiplication is done first.*
$= 35$

2. Calculate $25-(2+3)\times 4$.

$25-(2+3)\times 4$ *The brackets are done first, $(2+3)=5$.*
$=25-5\times 4$ *This is multiplied by 4 next, giving 20.*
$=25-20$ *Lastly this is subtracted from 25.*
$=5$

3. Calculate $22+(3+5)^2\times 2$.

$22+(3+5)^2\times 2$ *Calculate what is inside the brackets first.*
$=22+8^2\times 2$ *Calculate any powers next.*
$=22+64\times 2$ *Work out the multiplication next.*
$=22+128$ *Lastly add the two numbers together.*
$=150$

Exercise 5.1

1. Work out the following.
   a. $4+3\times 2-1$
   b. $6\times 2+4\times 3$
   c. $3\times 5-2-7$
   d. $8+4\times 8-40$
   e. $6-2\times 3\times 4$
   f. $7\times (4+2)-3$
   g. $8+(6+3)\div 3$
   h. $16\div (2+2)^2+8$
   i. $4+4\times 4-4$
   j. $(4+4)^2\div (8-4)$

2. Two pupils, Carla and Ibrahim, are discussing the following calculation.
   3×2^2
   Carla says the answer is 12, whilst Ibrahim says the answer is 36.
   a. Which pupil is right? Give a **convincing** reason for your answer.
   b. What mistake is the pupil who got it wrong making?

3. In the following calculations, insert brackets in order to make them correct.
   a. $3+2\times 5+4=29$
   b. $3+2\times 5+4=21$

4. Insert any of the following symbols, $+, -, \times, \div$ or $(\ )$, in between the numbers to make the calculations correct.
   a.   8   6   4 $=10$
   b.   8   6   4 $=6$
   c.   7   6   3 $=9$
   d.   7   6   3 $=39$
   e.   7   6   3 $=21$

f	8	4	2		= 4
g	8	4	2		= 2
h	2	3	5	2	= 20
i	2	3	5	2	= 18
j	3	9	8	5	= 36

5. Explain whether the use of brackets in the following calculations are necessary. Give a **convincing** reason for your answers.
   a $8 + (2 \times 6) = 20$
   b $(3 + 2)^2 - 6 = 19$
   c $(4^2 - 6) + 8 = 18$

6. The following questions all contain four 4's as shown below.

       4   4   4   4

   a i) Insert any of the following signs, +, −, ×, ÷ or ( ), in between the numbers to make the following calculation correct.

       4   4   4   4   = 1

      ii) Can you find a different way of getting an answer of 1 using four 4's?

   b Insert any of the following signs, +, −, ×, ÷ or ( ), in between the numbers to make the following calculation correct.

       4   4   4   4   = 2

   c i) Using just four 4's each time and any mathematical operation that you know of, can you make each of the answers from 3 to 20? Check your answers using a calculator.

      ii) Working with a partner, try to find more than one way to calculate each of the numbers 1–20. Check your answers using a calculator.

LET'S TALK

For part (c), with a partner decide on the different mathematical operations that you will allow.

Decide also whether you will allow two of the 4's to be written as 44.

Now you have completed Unit 5, you may like to try the Unit 5 online knowledge test if you are using the Boost eBook.

6 Algebra beginnings – using letters for unknown numbers

- Understand that letters can be used to represent unknown numbers, variables or constants.
- Understand that the laws of arithmetic and order of operations apply to algebraic terms and expressions.

In **algebra** letters are used to represent unknown numbers. Often the task is to find the value of the unknown number, but not always. This unit introduces the main forms that algebra can take.

Expressions

An expression is used to represent a value in algebraic form. For example,

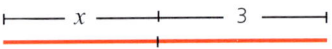

The length of the line is given by the expression $x+3$. Here we are not being asked to find the value of the unknown value x because the total length of the line is not given.

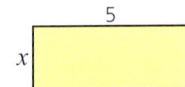

The perimeter of the rectangle is given by the expression $x+5+x+5$ which can be simplified to $2x+10$.

The area of the rectangle is given by the expression $5x$.

In the examples above, x, $2x$ and $5x$ are called **terms** in the expressions.

The numbers in front of the x in each case are called **coefficients**, i.e. in the term $5x$ the '5' is the coefficient.

An expression is different from an **equation**. An **equation** contains an equals sign (=), which shows that the expressions either side of it are equal to each other.

For example, the equation

$x+1=y-2$

tells us that the expressions $x+1$ and $y-2$ are equal to each other.

> **KEY INFORMATION**
> Note that $5x$ means 5 times x.
>
> When multiplying in algebra the × sign is not used.

6 Algebra beginnings – using letters for unknown numbers

In the earlier rectangle, if we are told that the area of the rectangle is 20 cm², then the equation $5x=20$ can be formed.

If we are asked to **solve** the equation, then we have to find the value of x that makes the left-hand side of the equation equal to the right-hand side.

> **Worked example**
>
> A boy is y years old. His sister is 4 years older than him, his younger brother 2 years younger than him and his grandmother 8 times older than him. Write an expression for the ages of his sister, younger brother and grandmother.
>
> As his sister is 4 years older than him, the expression for her age is $y+4$.
>
> As his younger brother is 2 years younger than him, his age is $y-2$.
>
> As the grandmother is 8 times older than him, her age can be expressed as $8y$.

Exercise 6.1

1. The lines below have the lengths shown. In each case write an expression for the length of the line.

   a ⊢— 7 —⊢— x —⊣   c ⊢— $2x$ —⊢— y —⊣

   b ⊢— $2x$ —⊢— 1 —⊣   d ⊢— $4x$ —⊢— $2y$ —⊢— 3 —⊣

2. i) Write an expression for the distance around the edge of each of these shapes.
   ii) Simplify your expression where possible.

   a  (rectangle: 4 by x)

   b  (rectangle: y by x)

   c 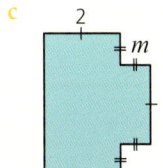 (shape with 2, m)

   d 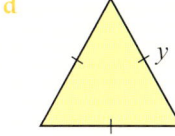 (triangle with y)

   e 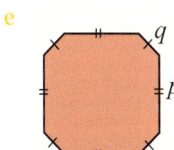 (octagonal shape with p, q)

   f 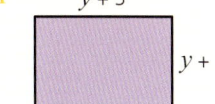 (rectangle: $y+3$ by $y+2$)

SECTION 1

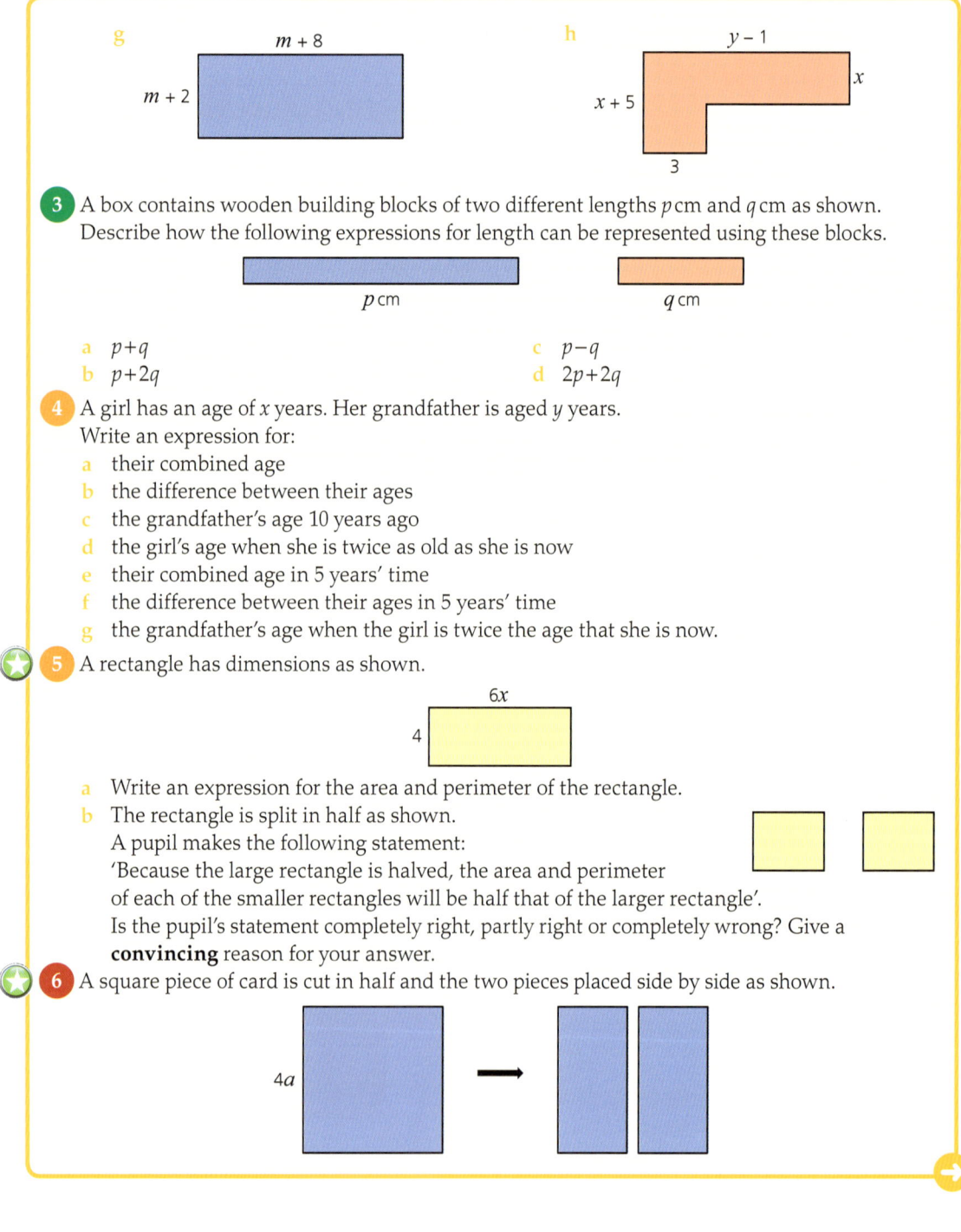

g m + 8, m + 2

h y − 1, x, x + 5, 3

3 A box contains wooden building blocks of two different lengths p cm and q cm as shown. Describe how the following expressions for length can be represented using these blocks.

p cm q cm

a $p+q$
b $p+2q$
c $p-q$
d $2p+2q$

4 A girl has an age of x years. Her grandfather is aged y years.
Write an expression for:
a their combined age
b the difference between their ages
c the grandfather's age 10 years ago
d the girl's age when she is twice as old as she is now
e their combined age in 5 years' time
f the difference between their ages in 5 years' time
g the grandfather's age when the girl is twice the age that she is now.

5 A rectangle has dimensions as shown.

6x, 4

a Write an expression for the area and perimeter of the rectangle.
b The rectangle is split in half as shown.
A pupil makes the following statement:
'Because the large rectangle is halved, the area and perimeter of each of the smaller rectangles will be half that of the larger rectangle'.
Is the pupil's statement completely right, partly right or completely wrong? Give a **convincing** reason for your answer.

6 A square piece of card is cut in half and the two pieces placed side by side as shown.

4a

a If the large square has a side length of 4a, write an expression for the total perimeter of the two rectangles.
b One of the two rectangles is halved lengthways and the pieces are also placed side by side.

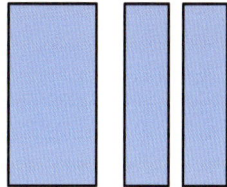

Write an expression for the total perimeter of the three rectangles.
c The three pieces are then arranged in different ways with no gaps between them as shown.

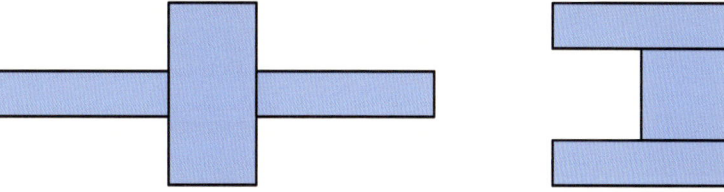

Explain **convincingly**, using algebra, why the perimeter of the two arrangements is not the same.
d Write an expression for the difference in the perimeters of the two arrangements above.

Order of operations when simplifying expressions

In Unit 5 you saw that calculations need to be carried out in a particular order. This order is not necessarily from left to right.

For example, the calculation 2+3×4 has the answer 14 (rather than 20) because the multiplication is done before the addition.

The order in which operations are carried out is as follows:

Brackets

Indices

Division/**M**ultiplication

Addition/**S**ubtraction

A useful way of remembering the order is with the shorthand **BIDMAS**.

The same order of operations applies when working with algebraic expressions.

SECTION 1

> The multiplication $3 \times 4a$ is done first.

Worked example

Simplify the expression $2a+3\times 4a-a$.
$2a+3\times 4a-a$
$=2a+12a-a$
$=13a$

Exercise 6.2

1 Simplify the following expressions.
  a $2a+3a-4a$
  b $2b-5b+7b-b$
  c $3c+2b+2c-4b$
  d $5d-4f+3e-2d+f-2e$

2 Write an expression for the area of each of the following shapes.

  a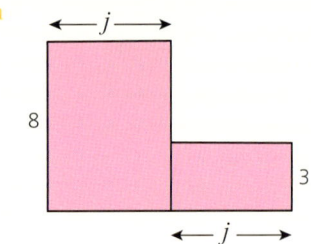
  b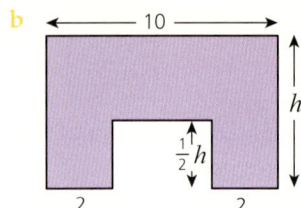

3 A compound shape is given opposite.
  a   Show that the total perimeter of the shape is given by the expression $6x+12$.
  b   Show that the total area of the shape is given by the expression $10x-6$.
  c   A pupil says that he has worked out the total area to be $12+6x-10+4x-8$.
      Explain whether this answer is correct.

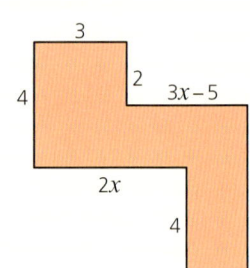

Substitution into expressions and formulae

KEY INFORMATION
The plural of formula is formulae.

$3p$ is an **expression**.

$3p=s$ is a **formula**.

$3p=9$ is an **equation**.

It is important that you understand the differences between these.

LET'S TALK
Can you think of any formulae you have used in your maths or science lessons?

An **expression** is just an algebraic statement.

6 Algebra beginnings – using letters for unknown numbers

KEY INFORMATION
A variable can take several values.

A **formula** describes a relationship between different **variables** and can be used to calculate values. If $p=5$, you can use the formula $3p=s$ to work out that the value of s is 15. If $p=8$, then $s=24$, and so on.

An **equation**, however, is only true for specific values of the variable. The equation $3p=9$ is only true when $p=3$.

Numbers can be **substituted** for the letters in both expressions and formulae.

Substitution into expressions

Worked example

Evaluate (work out) the expressions below when $a=3$, $b=4$ and $c=5$.

a $2a+3b-c$
  $=(2\times3)+(3\times4)-(5)$ *The multiplications are done first,*
  $=6+12-5$ *then the addition and subtraction.*
  $=13$

b $3ac+\dfrac{b}{2}=(3\times3\times5)+\dfrac{4}{2}$ *The multiplications and division are done first*
  $=45+2$ *then the addition.*
  $=47$

Exercise 6.3

Evaluate the expressions in questions 1 and 2 when $p=2$, $q=3$ and $r=5$.

1 a $3p+2q$ b $4p-3q$ c $p-q-r$ d $3p-2q+r$

2 a $pq+\dfrac{10}{r}$ b $\dfrac{pq+2p}{r}$ c $\dfrac{24}{q}+p\times r$ d $\dfrac{p}{4}\times 2r-\dfrac{4q}{3}$

Substitution into formulae

The perimeter, P, of a rectangle is the distance around it.

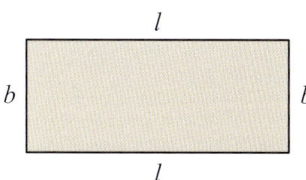

For this rectangle the perimeter is

$l+b+l+b$

or $\quad 2l+2b$

This can be written as:

$P=2l+2b$

We obtained this formula from the diagram. This is one way to **derive** a formula.

The area, A, of the rectangle can be written as:

$A=l\times b=lb$

Worked example

Using the formula $P=2l+2b$, calculate the perimeter of a rectangle if $l=3$ cm and $b=5$ cm.

$P=2\times3+2\times5$

$P=6+10$

$P=16$ cm

Exercise 6.4

1. Calculate the perimeter and area of each of these rectangles of length l and breadth b. Write the units of your answers clearly.
   - a $l=4$ cm, $b=7$ cm
   - b $l=8$ cm, $b=12$ cm
   - c $l=4.5$ cm, $b=2$ cm
   - d $l=8$ cm, $b=2.25$ cm
   - e $l=0.8$ cm, $b=40$ cm
   - f $l=1.2$ cm, $b=0.5$ cm
   - g $l=45$ cm, $b=1$ m
   - h $l=5.8$ m, $b=50$ cm

2. Marta and Raul are studying the two rectangles A and B below.

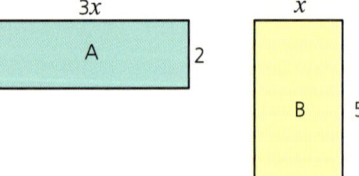

   a They calculate that the perimeter of A is given by the formula $P=6x+4$ and that the perimeter of B is given by the formula $P=2x+10$. Show that they are correct.

6 Algebra beginnings – using letters for unknown numbers

> **KEY INFORMATION**
> There are 1000 milliamps in 1 amp.
>
> In other words, 1 milliamp is 1/1000th of an amp.

> **LET'S TALK**
> What other examples use 'milli' to represent 1/1000th of another measure?

  b Marta says that the perimeter of A will **always** be greater than the perimeter of B because it has a '$6x$' in its formula which is bigger than '$2x$'. Raul disagrees. He says that the perimeter of B will **always** be greater than that of A because it has a '+10' in its formula which is bigger than '+4'. Explain whether either of them is correct.

3 In physics this formula is used in calculations about electricity:
   $V = IR$
   V is the voltage in a circuit in volts
   I is the current in amps
   R is the resistance in ohms.
   Without using a calculator, calculate the voltage V when
   a $I = 7$ amps, $R = 60$ ohms
   b $I = 8$ amps, $R = 400$ ohms
   c $I = 0.3$ amps, $R = 2000$ ohms
   d $I = 80$ milliamps, $R = 5000$ ohms

▶ Now you have completed Unit 6, you may like to try the Unit 6 online knowledge test if you are using the Boost eBook.

Organising and presenting data

- Record, organise and represent categorical, discrete and continuous data. Choose and explain which representation to use in a given situation.

Organising data

In Unit 3 of this book you encountered different types of data. These included **discrete**, **continuous** and **categorical** data. Data can be organised in several ways; which method is chosen depends largely on the type of data being collected. This unit introduces you to the main types of diagram used for organising and presenting data.

Constructing a **tally** and **frequency table** is a simple way of recording the number of results in each category.

For example, a survey is carried out to test the manufacturer's claim that there are 'about 36 chocolate buttons in each packet'.

The number of buttons in each of 25 packets is counted, giving the figures below.

35	36	34	37	36	36	38	37	36	35	38
34	35	36	36	34	37	38	37	36	35	36
36	37	36

Displayed as a list, the numbers are not clear.

However, they are easier to analyse if they are recorded in a tally and frequency chart like this.

> **LET'S TALK**
> Why is a diagonal line used for every fifth tally mark?

Number	Tally	Frequency
34	III	3
35	IIII	4
36	IIII IIII	10
37	IIII	5
38	III	3

The tally column is filled in as the survey is being carried out.
The frequency column is completed by counting up the tally marks at the end of the survey.

7 Organising and presenting data

Sometimes, if there is a big range in the data, it is more useful to group the data in a **grouped frequency table**. The groups are chosen so that no data item can appear in two groups.

For example, the ages of 30 residents in a care home are shown below.

98	71	76	77	72	78	77	73	76	86
75	79	81	105	100	74	82	88	91	96
85	90	97	102	83	101	83	84	80	95

Constructing a tally and frequency table with a list of individual ages will not be very useful as most ages in the range will only have one or two results. Grouping the data into the age ranges 71–80, 81–90 etc. produces this more useful table:

Age	Tally	Frequency
71–80	|||| |||| ||	12
81–90	|||| ||||	9
91–100	||||		6
101–110	|||	3

The ages could have been grouped 71–75, 76–80, 81–85 etc. The group size is the decision of the person collecting the data, but it is important that the groups are all the same size and do not overlap.

The examples above were for discrete data. However tally and frequency tables can be used for organising continuous data too.

The table below shows the time taken to run 100 m by a class of 30 children.

Time (seconds)	Tally	Frequency
12–13	||	2
13–14	||||	4
14–15	|||| |||	8
15–16	|||| ||||		11
16–17	||||	4
17–18			1

It seems that some numbers appear in more than one group. However with continuous data, results can take any value. A pupil may complete the race in 14.9 seconds, but a more accurate stopwatch may record the time as 14.99 seconds or even

14.999 seconds. So the group 14–15 for example is taken to mean any time from 14 up to but not including 15 seconds. The group 15–16 is taken to mean any time from 15 up to but not including 16 seconds and so on.

So far the examples given have only dealt with one variable. In the first example the variable was the number of chocolate buttons in a pack, in the second example the variable was the age of residents in a care home, whilst in the final example it was the time taken to run 100 m. If the data for two variables are being collected, an effective method is to use a **two-way table**.

> **Worked example**
>
> The maths test results of a group of 40 pupils are shown in the two-way table below:
>
> | | Test results % | | | | |
> |---|---|---|---|---|---|
> | | 1–20 | 21–40 | 41–60 | 61–80 | 81–100 |
> | **Boys** | 1 | 4 | 6 | 5 | 2 |
> | **Girls** | 0 | 5 | 7 | 8 | 1 |
>
> a What are the two variables?
>   Test results % and gender
> b How many girls sat the test?
> $5+7+8+1=21$ 21 girls sat the test.
> c To pass the test, pupils have to get more than 60% in the test. How many pupils failed the test?
> $1+4+6+5+7=23$ 23 pupils failed the test.

7 Organising and presenting data

Exercise 7.1

1 The midday temperature (°C) of all the days in July at a holiday resort are recorded and the results shown in the frequency table below.

Temperature (°C)	26	27	28	29	30	31	32
Frequency	1		8	6	7	2	4

a How many days had a temperature of at least 29°C?
b How many days had a temperature of 27°C?

2 The lengths (cm) of 50 worms of the same species are measured and recorded.
The first 45 results are shown in the grouped frequency table below.

Length (cm)	Tally	Frequency
10–12	⋕ I	
12–14	⋕ ⋕ III	
14–16		17
16–18	⋕	
18–20		4

a Explain whether the length of a worm, measured as accurately as possible, is an example of discrete or continuous data. Justify your answer.
b The remaining five worms were measured and their lengths recorded as:
10.6 cm   16.1 cm   13.9 cm   12.0 cm   11.9 cm
Copy and complete the grouped tally and frequency table to include the remaining five worms.
c It is known that females of this species of worm always have length in the range 10–16 cm, whilst males of the species always have lengths in the range 14–20 cm.
   i) How many female worms were definitely part of the data collection? Justify your answer.
   ii) What is the maximum number of male worms measured as part of this survey?

3 A survey is carried out amongst 40 children in a school to see what is their favourite animal. The results are presented in the two-way table below.

	Lion	Camel	Monkey	Mouse	Other	Total
Boys	5	1		2	6	24
Girls		4		5	0	
Total	7					40

a Copy and complete the table.
b How many girls said lions were their favourite animal?
c How many boys chose monkey as their favourite animal?
d If a pupil said that a rabbit was their favourite animal, decide whether it was a boy or girl who answered. Give a **convincing** reason for your answer.

SECTION 1

Displaying data

Once the data has been collected, it can be displayed in several ways. Which method is chosen depends on the type of data collected and the audience it is intended for.

One of the simplest and most effective is to use a **Venn diagram**.

A Venn diagram arranges the data into categories as shown in the example below.

Worked example

The numbers 1–20 are grouped into those that are multiples of two and those that are multiples of three as shown in the Venn diagram.

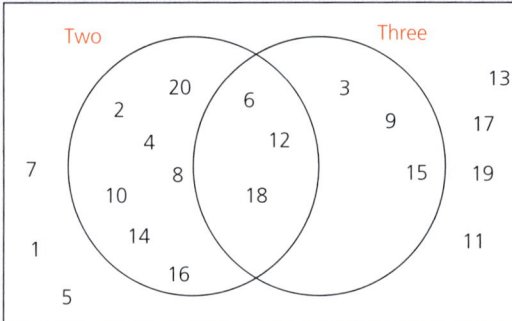

All the numbers 1–20 appear within the rectangle. Those inside the circle labelled '**Two**' are multiples of two and those inside the circle labelled '**Three**' are multiples of three.

a Which numbers are multiples of two and three?

The numbers 6, 12 and 18.

> They appear in the overlap of the two circles.

b Which numbers are multiples of neither two nor three?

The numbers 1, 5, 7, 11, 13, 17 and 19.

> They appear outside the circles, but still within the rectangle.

Another common way or arranging data into categories is to use a **Carroll diagram**.

7 Organising and presenting data

Worked example

The numbers 1–20 are arranged into four categories. These are multiples of 3, not multiples of 3, even numbers and odd numbers. In a Carroll diagram they can be arranged as shown.

	Even	Odd
Multiple of 3	6, 12, 18	3, 9, 15
Not a multiple of 3	2, 4, 8, 10, 14, 16, 20	1, 5, 7, 11, 13, 17, 19

a   Which of the numbers are odd multiples of 3?

3, 9 and 15 are odd multiples of 3.

b   Which numbers are even multiples of 3?

6, 12 and 18 are all even multiples of 3.

Exercise 7.2

1 In a school sports day, pupils can enter track events (T) and field events (F). The Venn diagram shows the number of pupils taking part in each type of event.

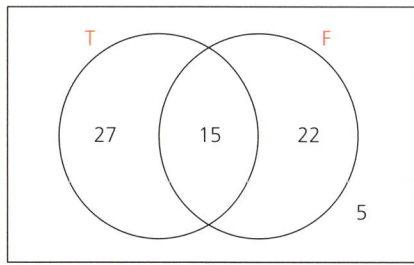

a   How many pupils entered track events?
b   How many pupils only entered field events?
c   How many pupils did not take part in any track or field events?

2 The members of a book group are asked whether they like fiction (F) or non-fiction (N) literature. The results are shown in the Venn diagram.

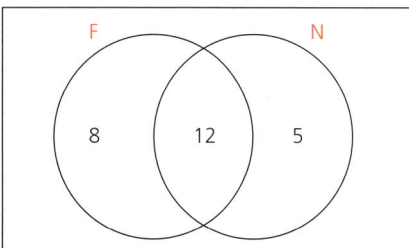

SECTION 1

a 20 people like fiction and 17 like non-fiction. One member says that therefore there must be 37 people in the book group. Explain whether this is correct or not. Justify your answer.
b How many members like just one of the types of literature?

3 Ten shapes are shown below.

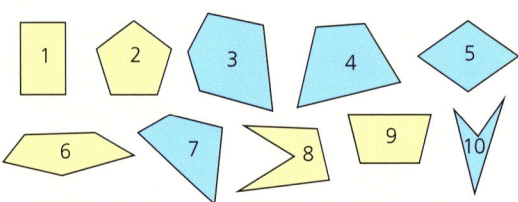

a Paul decides to classify the shapes in a Carroll diagram and draws one as shown:

	4-sided	Blue
5-sided		
Yellow		

Explain why Paul's Carroll diagram is not suitable for organising the 10 shapes.
b Design a Carroll diagram which will be suitable for **classifying** the shapes and enter the numbers of the shapes in the correct part of your diagram.

Probably the most common way of displaying data is the **bar graph** or **frequency diagram**. It is quick and easy to draw, and straightforward to understand.

Worked example

120 students from years 7, 8 and 9 of a school carry out a survey to see which subjects are most popular. Their results are shown in the frequency table.

Subject	Sport	Science	Maths	Art	Languages	**Total**
Frequency	40	20	30	15	15	**120**

Show this information on a frequency diagram.

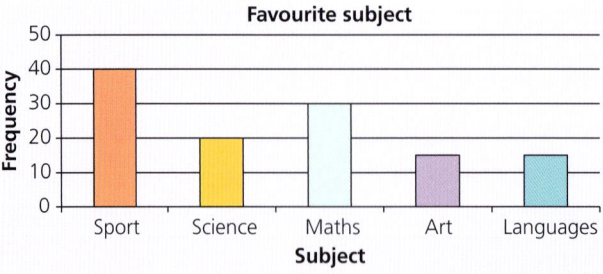

7 Organising and presenting data

The graph is fully labelled.
The bars are all the same width. As the data is categorical the bars do not touch.
The height of each bar represents the frequency.

A breakdown of the number of girls and boys voting for their favourite subject is shown below.

Subject	Sport	Science	Maths	Art	Languages	Total
Frequency (Girls)	16	10	18	5	6	**55**
Frequency (Boys)	24	10	12	10	9	**65**
Total	**40**	**20**	**30**	**15**	**15**	**120**

If the results of the girls' responses are to be compared with the boys' responses, rather than draw two separate frequency diagrams, it is more useful to plot a dual frequency diagram.

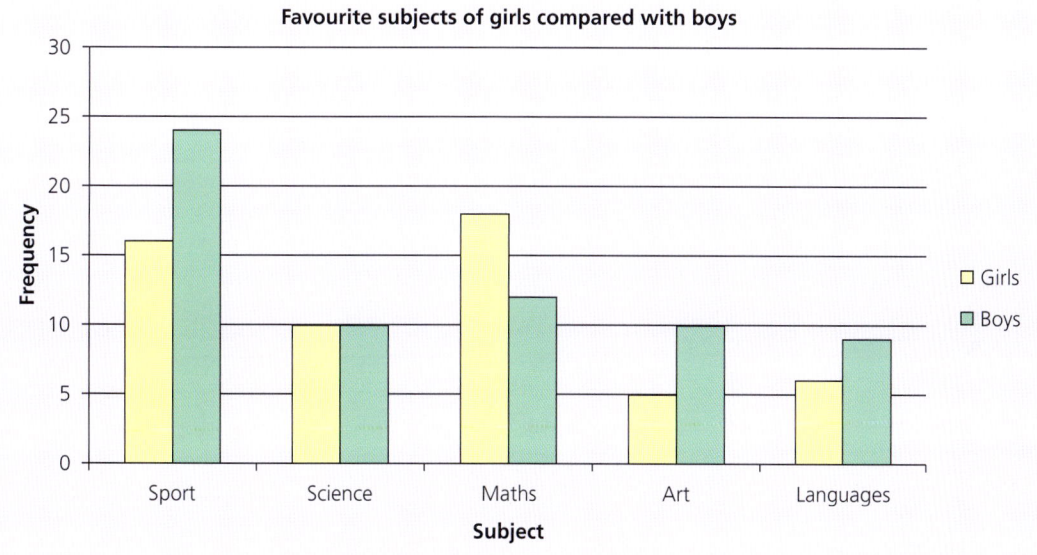

The graph is fully labelled.
The bars are all the same width. The girls' columns are presented next to the boys' columns within each category. This makes comparisons between the two easier.

The two types of bar chart shown so far can be combined to form a **compound bar chart**. As shown on the next page, each subject consists of one bar, but each bar is divided into girls and boys.

SECTION 1

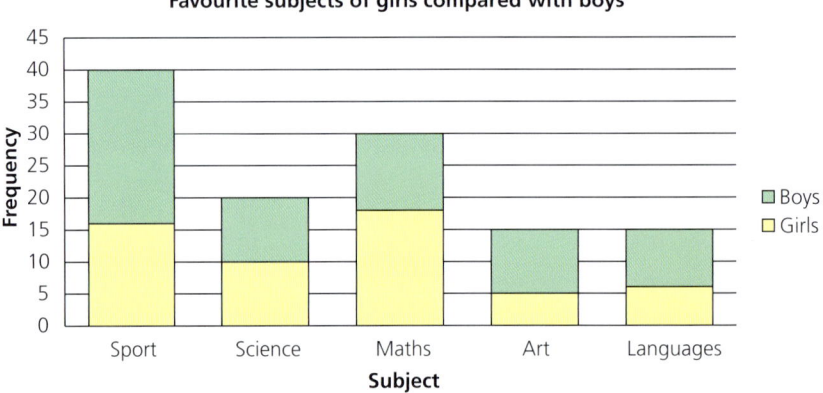

Frequency diagrams can also be used to display discrete grouped data and continuous grouped data. The difference between the two is that with discrete data the bars do not touch, whilst with continuous data they do.

Worked example

A train company wants to see how late its trains are. It conducts a survey of 100 trains and the results are shown in the grouped frequency table and grouped frequency diagram.

Lateness (mins)	Frequency
0–5	54
5–10	27
10–15	13
15–20	5
20–25	1

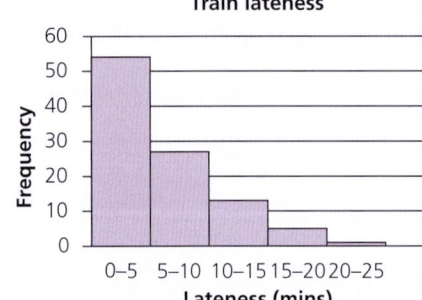

The graph shows that most trains are late between 0 and 5 mins.

Note that because the data is continuous, there are no gaps between the bars. 0–5 means from 0 up to but not including 5, 5–10 means from 5 up to but not including 10 etc.

7 Organising and presenting data

Exercise 7.3

1. Twelve people were asked which sandwiches they had bought from a sandwich shop. Their answers were:

   Chicken   Tuna   Egg   Chicken
   Egg   Tomato Chicken Tuna
   Tomato   Egg   Chicken Chicken

   Show this information on a bar chart.

2. Five different types of data collection are listed:

   Discrete   Continuous   Grouped discrete   Grouped continuous   Categorical

   a Give an example of a data collection for each of the types mentioned.
   b Collect your own data for one of the types listed and present the results in an appropriate frequency diagram. Justify your choice of diagram.
   c Collect your own data for a different type listed and present the results in another frequency diagram.

3. A sports club canteen keeps a record of the number of drinks it sells over two days. Part of the dual frequency graph is shown.

   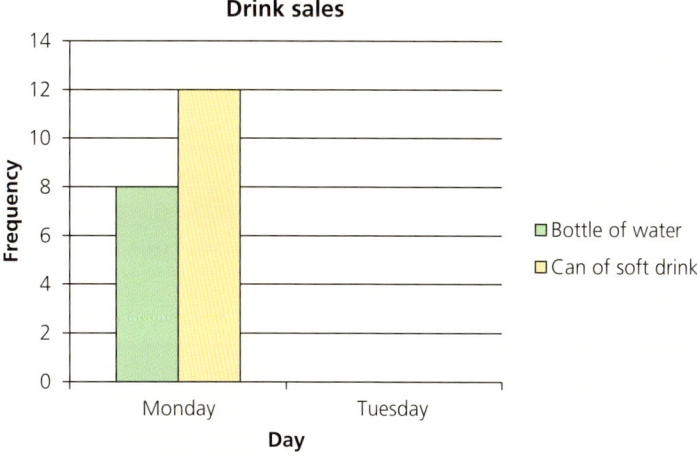

   Bottles of water are sold for $1.20 and cans of soft drink are sold for $1.60.
   a The total made from selling water bottles over the two days is $19.20. Calculate the number of bottles of water sold on Tuesday.
   b The total made from selling both types of drink over the two days is $46.40.
      Copy and complete the dual frequency diagram to show the drinks sold on Tuesday.

4. A fitness club carries out a survey to find out the ages of its members. Here are the results.

22	18	23	17	44	42	50	19	21	23	11	16
38	55	62	41	17	19	23	36	38	42	35	33
18	22	63	48	9	7	17	23	36	48	54	60

SECTION 1

a Make a grouped tally and frequency table using the age groups 1–10, 11–20, 21–30 etc.
b Is the data grouped discrete data or grouped continuous data? Justify your answer.
c Draw a frequency diagram of the data.
d What does the data tell you about the ages of people at the fitness club?

5 A zoologist weighs a sample of mice found in a field.
The mass (g) of 50 mice is recorded, grouped and shown in the two-way table below.

Mass (g)	Male	Female
60–80	1	4
80–100	5	10
100–120	▓	8
120–140	▓	▓

KEY INFORMATION
It is not always possible to survey a whole population so a **sample** is taken. The aim of a sample is to be **representative** of the population, i.e. show the same properties as the whole population.

Unfortunately the zoologist's pen leaked during the data collection.
a It is known that as a population 26% are males with a mass of 100–120 g and there are twice as many males as females with a mass of 120–140 g. Assuming the sample is representative of the whole population, copy and complete the table.
b Draw a compound grouped frequency diagram of the data.

In frequency diagrams and bar graphs, the frequency is represented by the **height** of the bar. Another popular way of displaying data is to use either a **pie chart** or a **waffle diagram**. On these, each frequency is represented by a **fraction of either a circle or a waffle diagram**.

Worked example

Look again at the data about students' favourite subjects. Show this information on a pie chart.
- First you need to express the frequency of each subject as a fraction of the total number of students.
  Sport is $\frac{40}{120} = \frac{1}{3}$ of the total,
  Science is $\frac{20}{120} = \frac{1}{6}$ of the total,
  Maths is $\frac{30}{120} = \frac{1}{4}$ of the total,
  and Art and Languages are $\frac{15}{120} = \frac{1}{8}$ each.
- To draw the pie chart without a protractor, an understanding of fractions helps.
  For example, Sport and Science together represent half of the total, and Maths, Art and Languages represent the other half of the total.

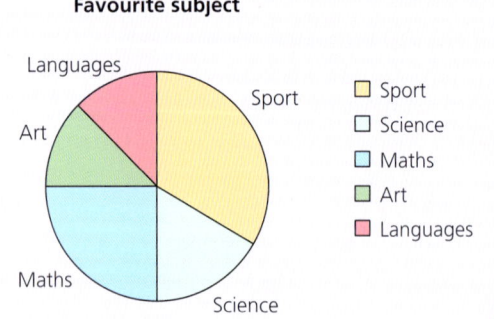

Favourite subject

7 Organising and presenting data

The pie chart has a heading and a key, and each slice is clearly labelled.

The pie chart is divided into slices, which are fractions of the circle.

The size of each slice represents the frequency, as a fraction of the total number of students.

A waffle is rectangular and split into squares. The number of squares that are shaded represent the fraction of the total. Therefore the example above can be represented in a waffle diagram which is split in to 120 squares.

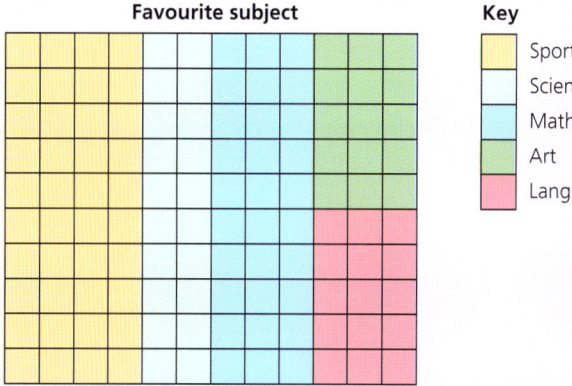

Although this particular waffle diagram has been split in to 120 squares, meaning 1 square represents 1 pupil, having the same number is not necessary. What is important is the fraction of the waffle diagram that is shaded.

Exercise 7.4

1. A baker's shop sells brown (B), white (W), wholemeal (M) and soda (S) bread. It keeps a record of the types of loaves it sells. The data is shown below.

B	B	W	M	W	M	B	W	M	S
B	B	B	W	M	W	M	B	M	B
S	W	W	B	B	W	M	M	M	M
B	M	M	W	M	W	B	S	M	M

a Construct a tally and frequency table of the results.
b Calculate the fraction of the total for each type of bread sold.
c Draw a waffle diagram to display this data.

2. The number of litres of milk consumed in a group of 160 houses per week are shown in the table.

Number of litres	1	2	3	4	5	6
Frequency	20	20	60	40	10	10

SECTION 1

    a Calculate the fraction of the total for each number of litres consumed.
    b Draw a pie chart to represent this data.
    c Draw a waffle diagram with 32 squares to represent this data.

3 90 students sat a maths exam. On the way out of the hall, they were asked whether they found it hard, OK or easy. Here are the responses expressed as a fraction of the total with one value missing.

Response	Easy	OK	Hard
Fraction		$\frac{1}{2}$	$\frac{1}{3}$

    a Show the results on a pie chart.
    b How many students found the exam easy? Show all your working clearly.

4 Four pie charts are shown below. Three of them can be paired with a waffle diagram which represents the same data.

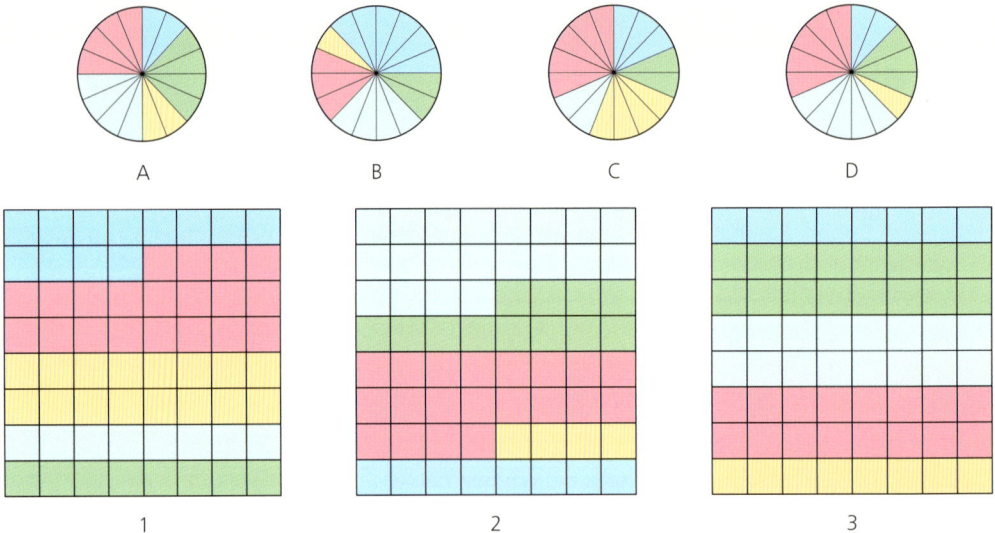

    a Match three of the pie charts with their corresponding waffle diagram.
    b Draw a waffle diagram to represent the remaining pie chart.

Line graphs, scatter graphs and infographics

This section will introduce you to other common types of graph.

A **line graph** is usually used to show a trend over a period of time.

7 Organising and presenting data

> **Worked example**
>
> The height (cm) of a sunflower is measured over a period of 8 weeks from when it starts to grow out of the ground. Measurements are taken on the same day each week.
>
> > **LET'S TALK**
> > Although measurements were only taken once a week, why is it acceptable to join the points?
>
>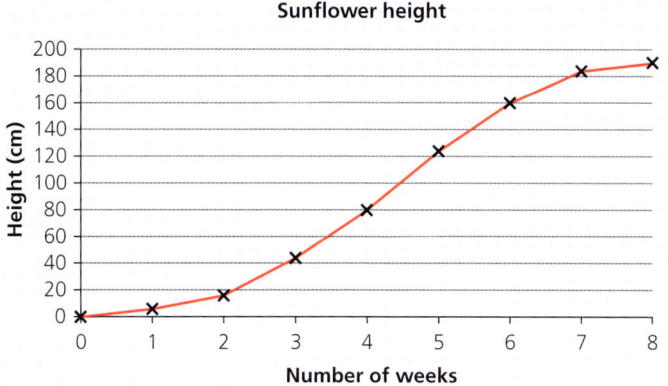
>
> The points are plotted and then joined up with straight lines. This is a useful graph as it is possible to make predictions of the height of the sunflower between measurements.
>
> Estimate the height of the sunflower after $4\frac{1}{2}$ weeks.
>
> A dashed line can be drawn up from $4\frac{1}{2}$ weeks until it meets the line of the graph. A horizontal line is then drawn until it meets the vertical axis as shown below.
>
> The height of the sunflower is approximately 100 cm after $4\frac{1}{2}$ weeks.
>
>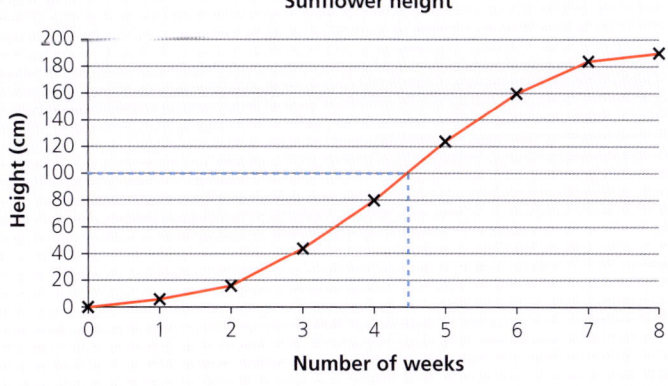

A **scatter graph** is used if you are trying to see if there is a relationship between two variables. The two variables are plotted as a single point on a graph.

SECTION 1

> **Worked example**
>
> A doctor decides to collect some data on the mass (kg) and height (cm) of toddlers attending her clinic. The average mass and height of toddlers of different ages are presented in the table below. The scatter graph is also plotted.
>
> | Age (months) | Average height (cm) | Average mass (kg) |
> |---|---|---|
> | 12 | 74.3 | 9.4 |
> | 13 | 75.3 | 9.7 |
> | 14 | 76.7 | 9.9 |
> | 15 | 78.1 | 10.1 |
> | 16 | 78.8 | 10.4 |
> | 17 | 80.0 | 10.6 |
> | 18 | 80.7 | 10.8 |
> | 19 | 81.8 | 11.0 |
> | 20 | 82.7 | 11.2 |
> | 21 | 83.5 | 11.6 |
> | 22 | 84.6 | 11.8 |
> | 23 | 85.2 | 12.0 |
>
>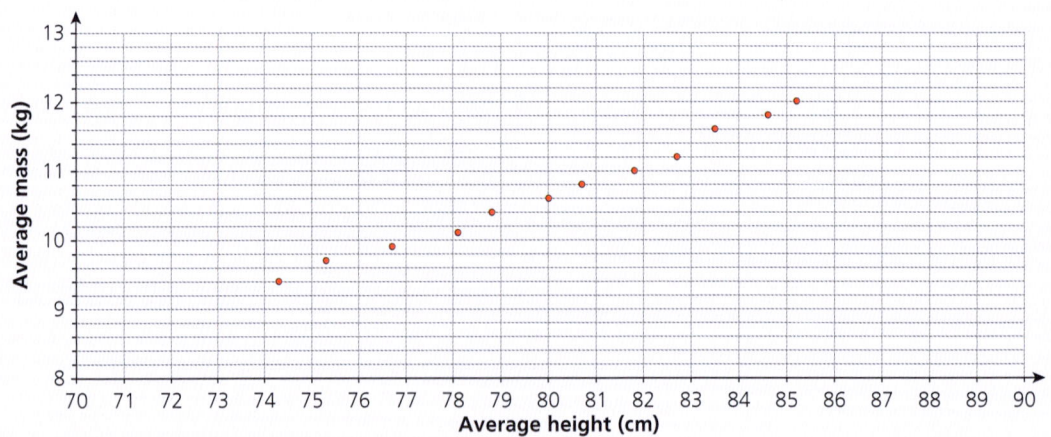
>
> Each point represents the average result for a particular age of toddler.
>
> The scatter graph helps us to decide whether there is a relationship between the two variables.
> Is there a relationship between the average height and average mass of a toddler?
> In this case as the points rise from left to right, it implies that taller toddlers tend to be heavier.

Infographics are used to display data in picture form. Quite often they are used to make the data more attractive and also to make it more easily understood by the viewer.

Worked example

The five tallest mountains in the world are ranked below.

Rank	Mountain	Height (m)
1	Everest	8848
2	K2	8611
3	Kangchenjunga	8586
4	Lhotse	8516
5	Makalu	8485

This data could be represented as a bar chart. However as an infographic it could be displayed as follows:

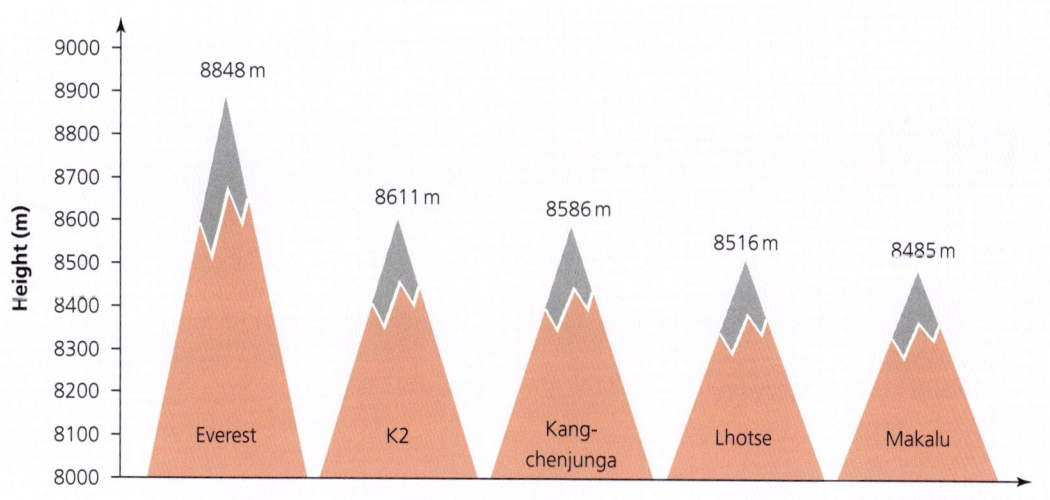

There is no right or wrong way of drawing an infographic. What is important is that the viewer can quickly see what the data is about. Here the use of mountains in the image makes it obvious that the data is to do with mountains.

SECTION 1

Exercise 7.5

  1 A coffee shop produces the following infographic:

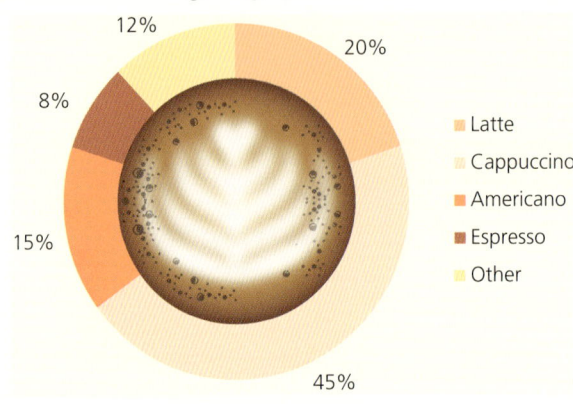

   a What do you think this infographic is showing?
   b Describe one thing that you like about this infographic.
   c Describe one thing that could be **improved**.
   d Design your own infographic for this data.

  2 A gardener measures the temperature (°C) in his greenhouse in two-hour intervals between 10:00 and 20:00 one day. The results are shown in the table below:

Time of day	10:00	12:00	14:00	16:00	18:00	20:00
Temperature (°C)	12	18	26	22	32	10

   a Plot a line graph of the data.
   b The gardener looks at the results and thinks that one of them must be wrong. Circle the result that is likely to be incorrect on your graph. Justify your answer.
   c On your graph mark a new position for the point you circled in part (b) and justify your answer.
   d Estimate the likely temperature of the greenhouse at 13:00.

7 Organising and presenting data

 3 Three scatter graphs are shown below.

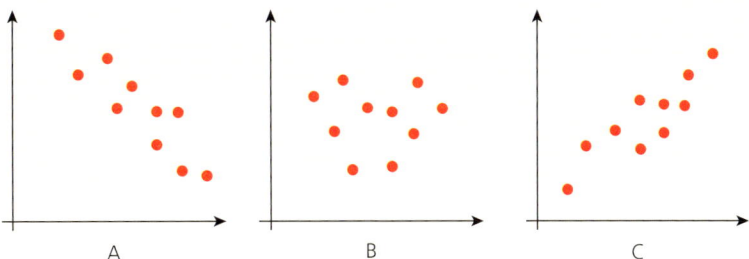

Data is collected for three different investigations. These are:
  i) the number of ice-creams sold per hour by an ice-cream vendor and the temperature during that hour
  ii) pupils' maths test results and their English test results
  iii) the value of a car and its age.
  a Match each scatter graph to the likely data from each investigation. Justify your choices.
  b Describe three different investigations which you think are likely to produce similar scatter graphs to the ones shown above. Justify your choices.

 Now you have completed Unit 7, you may like to try the Unit 7 online knowledge test if you are using the Boost eBook.

8 Properties of three-dimensional shapes

- Identify and describe the combination of properties that determine a specific 3D shape.
- Derive and use a formula for the volume of a cube or cuboid.
- Use knowledge of area and properties of cubes and cuboids to calculate their surface area.

Three-dimensional shapes

A three-dimensional shape is a solid figure or object with three dimensions, often described as length, width and height.

Some common three-dimensional shapes are:

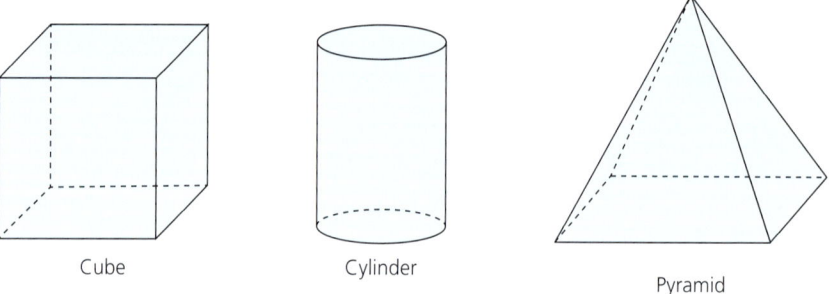

Cube Cylinder Pyramid

The different parts of a three-dimensional shape have specific mathematical names.

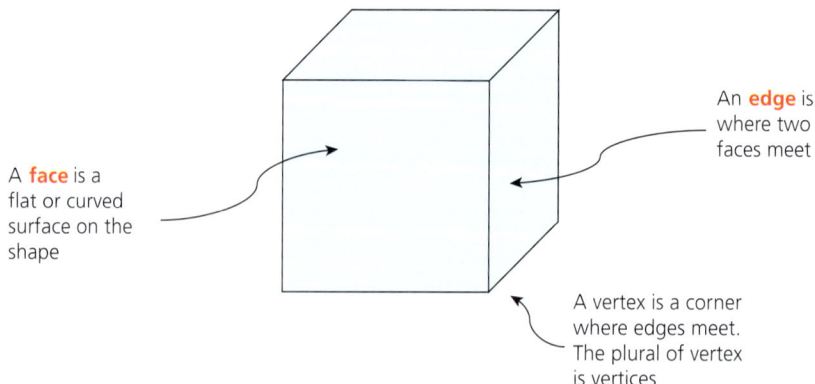

An **edge** is where two faces meet

A **face** is a flat or curved surface on the shape

A vertex is a corner where edges meet. The plural of vertex is vertices

8 Properties of three-dimensional shapes

Worked example

Count the number of faces, edges and vertices on each of the following shapes.

A cube

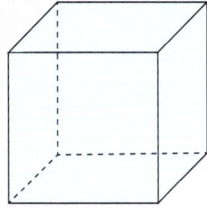

Number of faces = 6
Number of edges = 12
Number of vertices = 8

A cylinder

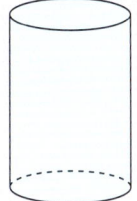

Number of faces = 3 (2 flat and 1 curved)
Number of edges = 2
Number of vertices = 0

Exercise 8.1

1. For the pyramids below count the number of faces, edges and vertices.

   a Square-based pyramid

   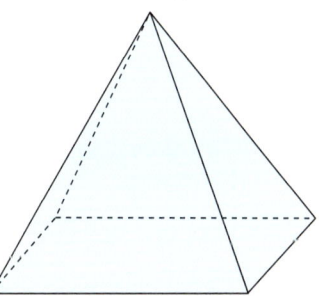

   b Hexagonal-based pyramid

   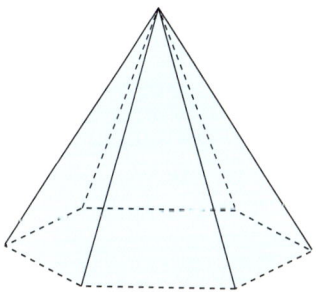

2. Sketch and name each of the three-dimensional shapes with the following properties:
   a Only one face and no edges or vertices.
   b Two faces, one edge and one vertex.
   c Five faces, nine edges and six vertices.

3. Two friends are discussing the properties of three-dimensional shapes. One states that if two different three-dimensional shapes have the same number of faces, then they must have the same number of edges. Prove **convincingly** that this is not true, by sketching two three-dimensional shapes with the same number of faces but with a different number of edges.

> **KEY INFORMATION**
> This is an example of **proof by counter example**, i.e. you find a case which contradicts the original statement.

55

SECTION 1

 4 Four different types of pyramid are shown below:

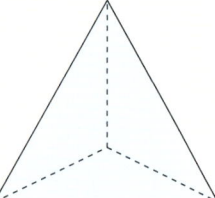 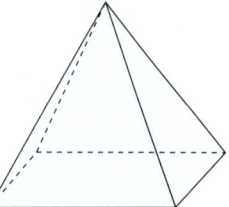 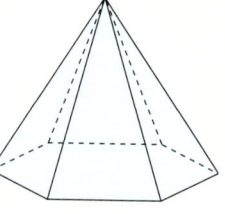 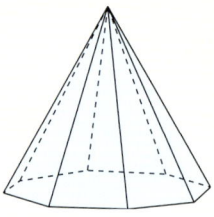

a It is stated that for every pyramid, there is always an even number of edges.
  i) Is this true for the four different pyramids above?
  ii) Explain whether you think the statement is always true. Give a **convincing** reason for your answer.

b It is also stated that every pyramid has the same number of faces as vertices.
  i) Is this true for the four different pyramids above?
  ii) Explain whether you think the statement is always true. Give a **convincing** reason for your answer.

> **LET'S TALK**
> With a friend, discuss your answers to these questions.

> **KEY INFORMATION**
> The units of volume include mm^3, cm^3 and m^3 for solid shapes.

> **LET'S TALK**
> All the units of volume mentioned so far use the 'cubed' notation ^{3}. Can you think of units of volume which don't use this?

Volume of a cuboid

The volume of a three-dimensional object or shape refers to the amount of space it occupies.

By taking a basic cube with side lengths of 1 cm, it is possible to work out the volume of other cubes and cuboids.

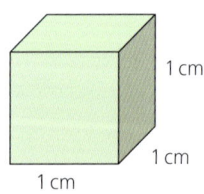

This cube has a volume of $1\,cm^3$.

By putting more of these cubes together, it is possible to form larger cubes and cuboids.

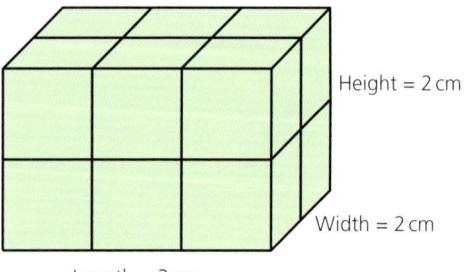

As this cuboid is made up of $12 \times 1\,cm^3$ it has a volume of $12\,cm^3$.

8 Properties of three-dimensional shapes

Exercise 8.2

1 In the following table several cuboids are drawn. Each cuboid is made from 1 cm³ cubes. For each cuboid write down the length, width and height and work out its volume.

	Cuboid	Length	Width	Height	Volume
a					
b					
c					
d					
e					

2 a Describe in words the relationship between the length, width and height of a cuboid and its volume.
  b Write the relationship you described in part (a) as a formula.

3 Calculate the volume of each of these cuboids, where L=length, W=width and H=height. Give your answers in cm³.
  a $L=4$ cm $W=2$ cm $H=3$ cm
  b $L=5$ cm $W=5$ cm $H=6$ cm
  c $L=10$ cm $W=10$ mm $H=4$ cm
  d $L=40$ cm $W=0.2$ m $H=50$ cm
  e $L=50$ mm $W=30$ cm $H=0.1$ m

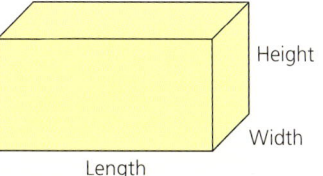

SECTION 1

4 This cuboid has a volume of 360 cm³. Calculate the length (in cm) of the edge marked x.

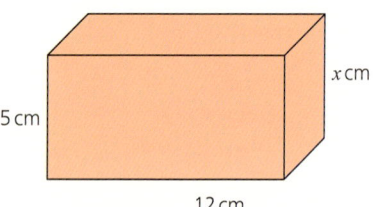

5 The volume of this cuboid is 180 cm³. Calculate the length (in cm) of the edge marked y.

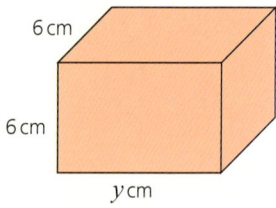

6 a This cuboid has volume 768 cm³ and the edges marked a are equal in length. Calculate the value of a.
   b In another cuboid of length 12 cm and volume 768 cm³, the width and height are not equal. Give a pair of possible values for their length.

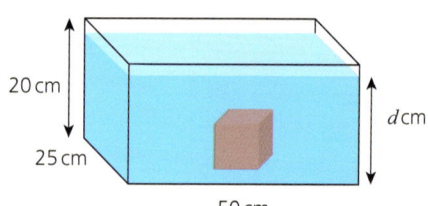

7 A tank of water in the shape of a cuboid has a length of 50 cm, a width of 25 cm and a height of 20 cm. The depth of water in the tank is 12 cm as shown below.

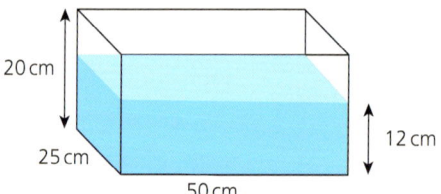

a A cube of side length 10 cm is placed at the bottom of the tank and the water level rises to a depth d cm as shown.
Calculate the depth of the water d cm.
b How many cubes can be placed in the tank before the water starts to spill over the top? Show all your working clearly.

Composite three-dimensional shapes

A **composite shape** is one which is made from other shapes. Here we will look at composite shapes which can be broken down into cubes and cuboids.

8 Properties of three-dimensional shapes

> **KEY INFORMATION**
> A prism is a 3D shape that, when sliced in a particular direction, is the same all the way through. It has a constant cross-section.

Worked example

Calculate the volume of this 'n'-shaped prism.

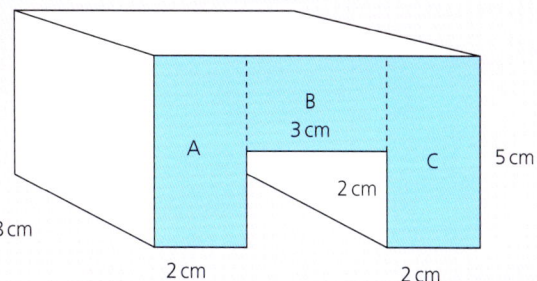

The shape can be broken down into three cuboids, labelled A, B and C. The volume of the prism is therefore the sum of the volumes of the three cuboids.

Volume of cuboid A = 2×8×5 = 80 cm³

Volume of cuboid B = 3×8×3 = 72 cm³

Volume of cuboid C = 2×8×5 = 80 cm³

Total volume of prism = 80 + 72 + 80 = 232 cm³

Exercise 8.3

Calculate the volume of the composite shapes in Questions 1–3.

1.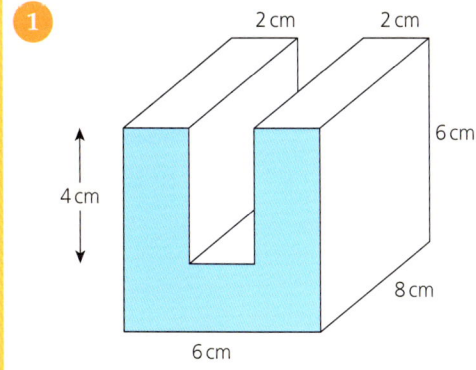

2.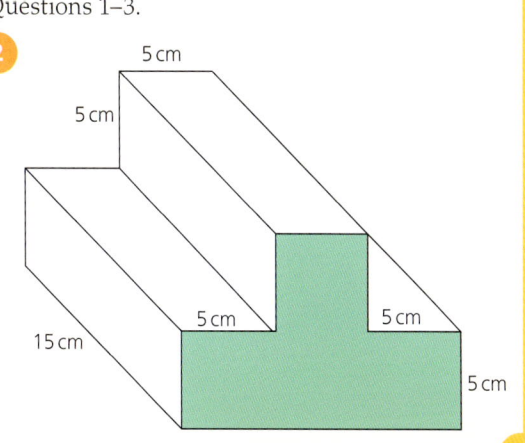

59

SECTION 1

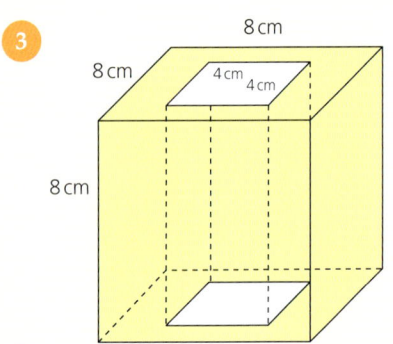

3

4 Design your own composite shape which can be split into two cuboids, with a total volume of 200 cm³.

5 Design your own composite shape which can be split into three cuboids, with a total volume of 500 cm³.

6 The cuboid below has dimensions as shown:

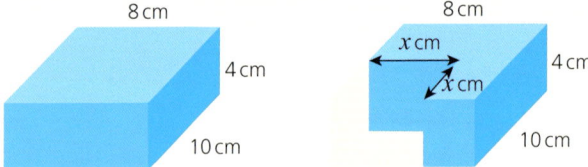

A cuboid with a depth of 4 cm but a length and width of x cm is cut out from one corner of the original cuboid as shown. The remaining shape has a volume of 199 cm³.
Calculate the value of x. Show all your working clearly.

Surface area of a cuboid

The surface area of a cuboid refers to the total area of the six faces of the cuboid. As each face is a rectangle, the total surface area involves finding the area of the six rectangles.

Worked example

Calculate the surface area of this cuboid.

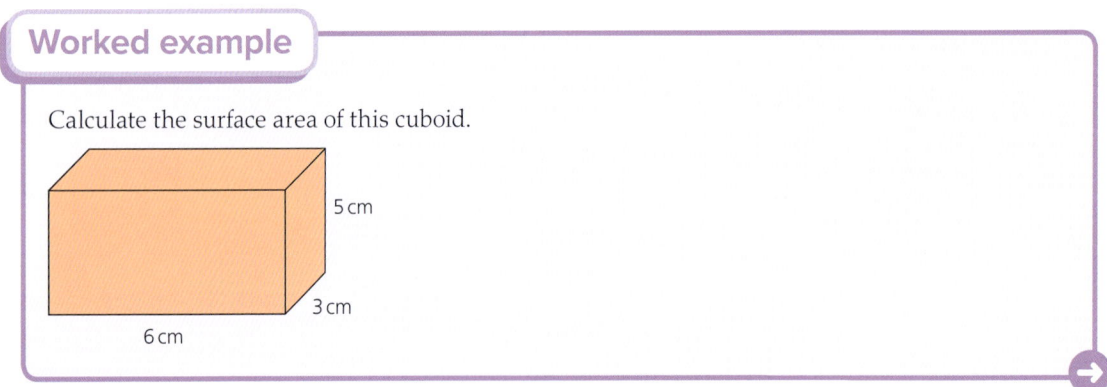

8 Properties of three-dimensional shapes

There are two ways of solving this problem.

Method 1: By calculating the area of each rectangular face:

As the front is the same as the back, the top the same as the bottom and the two sides equal to each other, the area can be worked out in pairs.

Area of front and back $= 6 \times 5 \times 2 = 60 \, cm^2$

Area of top and bottom $= 6 \times 3 \times 2 = 36 \, cm^2$

Area of both sides $= 3 \times 5 \times 2 = 30 \, cm^2$

Total surface area $= 60 + 36 + 30 = 126 \, cm^2$

Method 2: By calculating the area of the net of the cuboid.

The net of a cuboid is the two-dimensional shape which, when folded up, forms the cuboid.

For example:

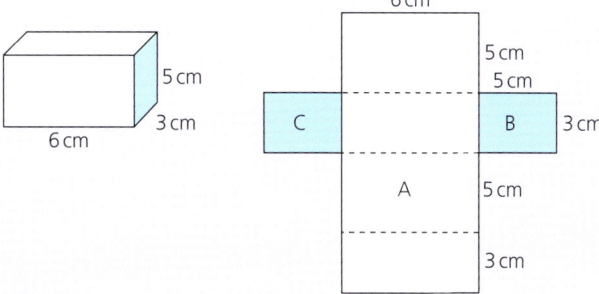

Area of large rectangle A $= 6 \times 16 = 96 \, cm^2$

Area of rectangle B $= 5 \times 3 = 15 \, cm^2$

Area of rectangle C $= 5 \times 3 = 15 \, cm^2$

Therefore total surface area $= 96 + 15 + 15 = 126 \, cm^2$

SECTION 1

Exercise 8.4

1. Calculate the surface area of the following cuboids.

   a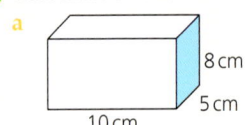

   b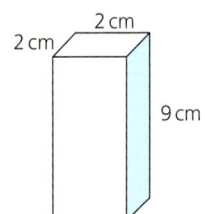

   c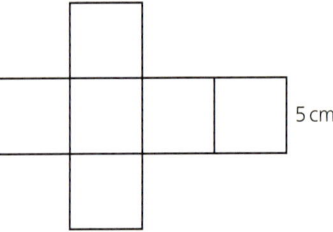

2. The net of a cube is shown below.

   Calculate
   a the surface area of the cube
   b the volume of the cube.

3. A cube has an edge length of x cm.
   a Show that the total surface area (A) can be calculated using the formula $A = 6x^2$.
   b Use the formula to calculate the total surface area of a cube of edge length 10 cm.

4. A cube has an edge length of 3 cm.
   a Calculate its total surface area.
   b If the edge length is doubled, how many times bigger does the surface area become?
   c If the edge length of the original cube is trebled, how many times bigger does the surface area become?
   d Predict, without calculating the surface area, how many times bigger the total surface area becomes if the edge length of the original cube is 10 times bigger. Justify your answer.

5 For the cuboid drawn below:
  a draw two possible nets for the cuboid
  b calculate the total surface area of both nets, showing clearly the dimensions of each part of the net.

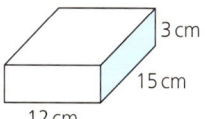

6 A room in the shape of a cuboid is shown below. The room has two identical square windows and a door with the dimensions given.

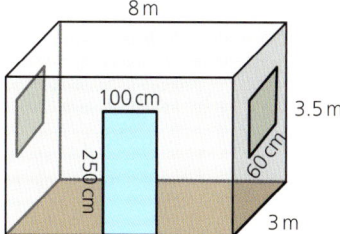
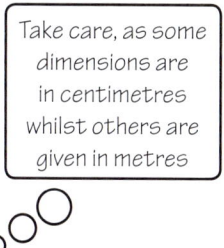

Take care, as some dimensions are in centimetres whilst others are given in metres

Vladimir wants to decorate the room, including the ceiling, with two coats of paint (he will not paint the floor, windows or door). If each paint pot claims it can cover 60 m² of surface area, how many pots will Vladimir need to buy? Show your working clearly.

7 Two students, Beatriz and Fatou, are discussing the relationship between a cuboid's volume and its total surface area. Beatriz states that cuboids with the same volume must have the same surface area. Fatou thinks Beatriz is wrong.
Which student is correct? Give a **convincing** reason for your answer.

Now you have completed Unit 8, you may like to try the Unit 8 online knowledge test if you are using the Boost eBook.

9 Multiples and factors

- Understand lowest common multiple and highest common factor (numbers less than 100).
- Use knowledge of tests of divisibility to find factors of numbers greater than 100.

Factors of a number are all the whole numbers which divide exactly into that number. For example, the factors of 12 are all the numbers which divide into 12 exactly. They are 1, 2, 3, 4, 6 and 12.

Multiples of a number are all the whole numbers which are in that number's times table. For example, the multiples of 3 are all the numbers in the 3× table. The first five are 3, 6, 9, 12 and 15, but there are in fact an infinite number of multiples of 3.

> **LET'S TALK**
>
> How can there be an infinite number of multiples of 3 if not all numbers are multiples of 3?
>
> Does this mean that there are more than an infinite number of numbers?

Highest common factors and lowest common multiples

The factors of 12 are 1, 2, 3, 4, ⓖ and 12.

The factors of 18 are 1, 2, 3, ⓖ, 9 and 18.

The **highest common factor (HCF)** of 12 and 18 is therefore 6, as it is the largest factor to appear in both groups.

The multiples of 6 are those numbers in the 6× table, i.e. 6, 12, 18, ㉔, 30 etc.

The multiples of 8 are those numbers in the 8× table, i.e. 8, 16, ㉔, 32, 40 etc.

The **lowest common multiple (LCM)** of 6 and 8 is therefore 24, as it is the smallest multiple to appear in both groups.

> **KEY INFORMATION**
>
> The highest common factor can also be called the greatest common divisor.

9 Multiples and factors

Exercise 9.1

1. Find the highest common factor of the following numbers:
   a. 8, 12
   b. 10, 25
   c. 12, 18, 24
   d. 15, 21, 27
   e. 36, 63, 108

2. Three cards each have a different factor of 18 written on them. Two of the numbers are shown, the third is hidden.
   If the number on the third card is not a multiple of 3, what must it be?

3. Find the lowest common multiple of the following numbers:
   a. 6, 14
   b. 4, 15
   c. 2, 7, 10
   d. 3, 9, 10
   e. 3, 7, 11

4. The lowest common multiple of two numbers is 60.
   a. What two numbers could they be?
   b. Is another pair of numbers possible? If so, what numbers are they?

5. The factors of 24 are arranged in a 3×3 grid, leaving one square blank. The totals of each row and column are shown below.

   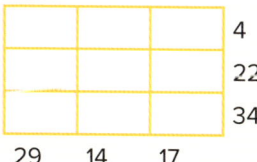

   a. Explain why the number 24 must appear in the bottom left square of the grid.
   b. Explain why the blank square must appear in the top row.
   c. Copy and complete the grid by inserting the factors of 24 in the correct squares.

Divisibility

It is useful to be able to check whether a number is divisible (can be divided) by any of the numbers from 2 to 10 without having to use a calculator.

There are quick methods that can be used to check for divisibility.

SECTION 1

Divisible by 2, 5 or 10

1	2	3	4	5	6	7	8	9	10
11	12	13	14	15	16	17	18	19	20
21	22	23	24	25	26	27	28	29	30
31	32	33	34	35	36	37	38	39	40
41	42	43	44	45	46	47	48	49	50
51	52	53	54	55	56	57	58	59	60
61	62	63	64	65	66	67	68	69	70
71	72	73	74	75	76	77	78	79	80
81	82	83	84	85	86	87	88	89	90
91	92	93	94	95	96	97	98	99	100

1	2	3	4	5	6	7	8	9	10
11	12	13	14	15	16	17	18	19	20
21	22	23	24	25	26	27	28	29	30
31	32	33	34	35	36	37	38	39	40
41	42	43	44	45	46	47	48	49	50
51	52	53	54	55	56	57	58	59	60
61	62	63	64	65	66	67	68	69	70
71	72	73	74	75	76	77	78	79	80
81	82	83	84	85	86	87	88	89	90
91	92	93	94	95	96	97	98	99	100

In the first grid all the numbers between 1 and 100 that are divisible by 2 have been shaded. We can generalise that multiples of 2 all end with either a 0, 2, 4, 6 or 8.

In the second grid all the numbers that are divisible by 5 have been shaded. All multiples of 5 end with either a 0 or 5.

Also, numbers divisible by 10 all end with a 0.

Divisible by 3 or 9

1	2	3	4	5	6	7	8	9	10
11	12	13	14	15	16	17	18	19	20
21	22	23	24	25	26	27	28	29	30
31	32	33	34	35	36	37	38	39	40
41	42	43	44	45	46	47	48	49	50
51	52	53	54	55	56	57	58	59	60
61	62	63	64	65	66	67	68	69	70
71	72	73	74	75	76	77	78	79	80
81	82	83	84	85	86	87	88	89	90
91	92	93	94	95	96	97	98	99	100

1	2	3	4	5	6	7	8	9	10
11	12	13	14	15	16	17	18	19	20
21	22	23	24	25	26	27	28	29	30
31	32	33	34	35	36	37	38	39	40
41	42	43	44	45	46	47	48	49	50
51	52	53	54	55	56	57	58	59	60
61	62	63	64	65	66	67	68	69	70
71	72	73	74	75	76	77	78	79	80
81	82	83	84	85	86	87	88	89	90
91	92	93	94	95	96	97	98	99	100

9 Multiples and factors

Adding the digits of each number together gives a total that is divisible by 3.

For example, 54 is divisible by 3 as 5+4=9 and 9 is divisible by 3.

This rule applies to larger numbers too:

372 is divisible by 3 as 3+7+2=12 and 12 is divisible by 3.

> Adding the digits of each number together gives a total that is divisible by 9.
>
> For example, 99 is divisible by 9 as 9+9=18 and 18 is divisible by 9.
>
> This rule applies to larger numbers too.
>
> 5877 is divisible by 9 as 5+8+7+7=27 and 27 is divisible by 9.

Divisible by 4, 6, 7 or 8

These are slightly trickier and you will need to be familiar with your times tables to use these rules efficiently.

A number is divisible by 4 if the last two digits of the number are divisible by 4.

For example, the number 1**28** is divisible by 4 as the last two digits are 28 and 28 is divisible by 4.

An alternative method would be to halve the number and then halve it again.

If the result is a whole number then the original number is divisible by four.

For example, the number 128 is divisible by 4 because:

128÷2=64 and 64÷2=32. As 32 is a whole number 128 is divisible by 4.

The number 246 is not divisible by 4 because:

246÷2=123 and 123÷2=61.5, which is not a whole number.

Similarly, a number is divisible by 8 if the last three digits of the number are divisible by 8.

For example, the number 2**432** is divisible by 8 as the last three digits are 432, and 432 is divisible by 8.

LET'S TALK

Why do we only need to consider the last two digits of the number to test whether it is divisible by 4?

LET'S TALK

Why do we only need to consider the last three digits of the number to test whether it is divisible by 8?

Hint: Is 1000 divisible by 8?

SECTION 1

You may need to do a quick division to realise this!

An alternative method would be to halve the number, halve it again and then again (i.e. halve the number three times).

If the result is a whole number then the original number is divisible by 8.

For example, the number 2432 is divisible by 8 because:

$2432 \div 2 = 1216$, $1216 \div 2 = 608$ and $608 \div 2 = 304$

As 304 is a whole number 2432 is divisible by 8.

For a number to be divisible by 6 it must be divisible by both 2 and 3, for example, 1104 is divisible by 6 as it is divisible by 2 (the number ends in a 4) and divisible by 3 ($1+1+0+4=6$ which is divisible by 3).

Divisibility by 7 involves the following steps:

For example, to test whether 1078 is divisible by 7:
- take the last digit off the number and double it ($8 \times 2 = 16$)
- subtract this from the remaining digits ($107 - 16 = 91$).

If the answer is divisible by 7, then the original number is too (91 is divisible by 7, therefore 1078 is divisible by 7).

> **KEY INFORMATION**
> To see why this method works, you may want to look it up on the internet.

Exercise 9.2

1. Copy the following table and tick the squares when the number is divisible by a number written along the top. One example has been started for you.

		2	3	4	5	6	7	8	9	10	25	100
a	50	✓			✓							
b	270											
c	1120											
d	135											
e	302400											

2. Four cards are arranged below to form a four-digit number.

   5 2 3 8

a Arrange the four cards so that the number is:
   i) divisible by 5
   ii) divisible by 9
   iii) divisible by 6
   iv) divisible by 8.
b In one of the questions in part (a) the order of the cards does not matter. Which one is it? Justify your answer.

3 Type this formula into cell A1 in a spreadsheet: =RANDBETWEEN(0,500).
This will generate a random integer (whole number) between 0 and 500.
Copy the formula down to cell A20 to generate 20 random numbers in the first column of the spreadsheet, for example as shown here.

	A	B
1	364	
2	293	
3	59	
4	112	
5	99	
6	19	
7	254	
8	275	
9	125	
10	276	
11	144	
12	300	
13	300	
14	267	
15	451	
16	318	
17	307	
18	95	
19	430	
20	352	

a Use divisibility tests to find out which of your random numbers are divisible by either 2, 3 or 6.
b Which of your random numbers are divisible by either 5, 10 or 100?
c Which of your random numbers are divisible by 9?
d Is it true that any number which is divisible by 9 is also divisible by 3? Explain your answer.
e Is it true that any number which is divisible by 3 is also divisible by 9? Explain your answer.

Now you have completed Unit 9, you may like to try the Unit 9 online knowledge test if you are using the Boost eBook.

Probability and the likelihood of events

- Use the language associated with probability and proportion to describe, compare, order and interpret the likelihood of outcomes.
- Understand and explain that probabilities range from 0 to 1, and can be represented as proper fractions, decimals and percentages.

 Probability is the study of **chance**, or the **likelihood** of an event happening.

In everyday language we use words that are associated with probability all the time. For example, "I *might* see that film", "I'm *definitely* going to win this race" or "It's *unlikely* that I'll pass this test". Some other words include:

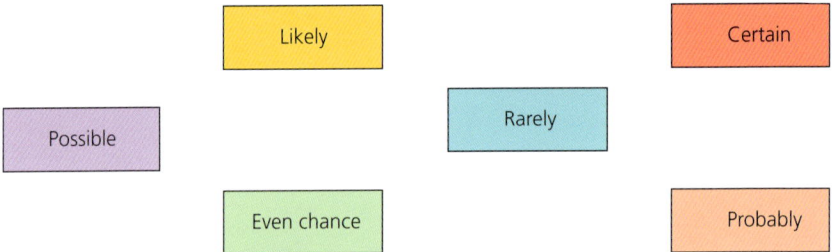

When using these words we are usually indicating how likely something is to happen. For example, an event that is 'likely' is more probable than an event that 'rarely' happens.

In this unit we will be looking at **theoretical probability**, that is, what you would expect to happen in theory. But, because probability is based on chance, what should happen in theory does not necessarily happen in practice.

With an ordinary coin, there are two **possible outcomes** (two results that could happen). These are heads or tails.

Each of these possible outcomes is an **equally likely outcome** if the coin is fair. This means that the coin is equally likely to land on heads or tails.

The probability of getting a head when the coin is flipped is $\frac{1}{2}$.

The probability is $\frac{1}{2}$ because getting a head is only one outcome out of two possible outcomes. This probability could also be written in decimal form as 0.5, or in percentage form as 50%.

10 Probability and the likelihood of events

> **Worked example**
>
> An **unbiased** spinner, numbered 1–4, is spun.
>   a i) Calculate the probability of getting a 2.
>
>   The probability of getting a 2 when the spinner is used is $\frac{1}{4}$.
>
>   ii) Write the probability of getting a 2 as a decimal and a percentage.
>
> $\frac{1}{4} = 0.25 = 25\%$
>
>   b Calculate the probability of getting a 7.
>
>   The probability of getting a 7 is $\frac{0}{4}$ as there is no number 7 on the spinner. (Note: $\frac{0}{4} = 0$)
>
>   Calculate the probability of getting either a 1, 2, 3 or 4.
>
>   The probability of getting a 1, 2, 3 or 4 is $\frac{4}{4}$ as there are four numbers out of four possible outcomes. It is certain that we will spin one of those numbers. (Note $\frac{4}{4} = 1$)
>
> **If an outcome has a probability of 0, it means the outcome is impossible.**
>
> **If an outcome has a probability of 1, it means the outcome is certain.**
>
> The probability of an event can be placed on a probability scale from 0 to 1 like this.
>
>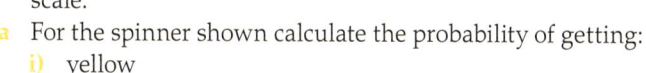

KEY INFORMATION
All probabilities lie in the range 0–1.

Exercise 10.1

1. a Write down at least 15 words which are used in everyday language to describe the likelihood of an event happening.
    b Draw a probability scale similar to the one above. Write each of your words from part (a) where you think they belong on the probability scale.

2. a For the spinner shown calculate the probability of getting:
    i) yellow
    ii) light blue
    iii) any blue.

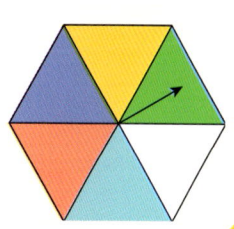

SECTION 1

b The spinner is spun 60 times.
  i) Estimate the number of times you would expect to get the colour red.
  ii) Would you definitely get red that number of times? Explain your answer.

3 Five cards are numbered with a different number from 1 to 10 as shown below. One card is covered.

> **LET'S TALK**
> 'At random' means that each choice is equally likely.

A card is chosen at random. What could be the number on the covered card if:
a the probability of picking an even number is 0.4
b the probability of picking a number less than five is 20%
c the probability of picking 10 is $\frac{1}{5}$?

4 The letters T, C and A can be written in several different orders.
a Write the letters in as many different orders as possible.
b If a computer writes these three letters in a random order, calculate the probability that:
  i) the letters will be written in alphabetical order
  ii) the letter T is written before both the letters A and C
  iii) the letter C is written after the letter A
  iv) the computer will spell the word CAT.

5 In one school, students have the option of studying either French (F) or Spanish (S) or neither. The incomplete Venn diagram below shows the number of students studying each language.

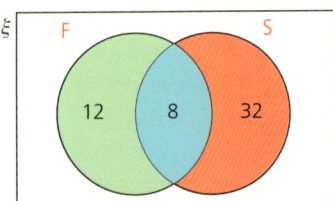

A student is chosen at random. If the probability of choosing a student who studies French is $\frac{1}{4}$, calculate:
a the total number of students in the year group
b the probability of choosing a student who studies neither language. Give your answer as a fraction, decimal and percentage.

6 500 balls numbered from 1 to 500 are placed in a large container. A ball is picked at random.
a Are the numbers an example of equally likely outcomes? Justify your answer.
b Calculate the probability that the ball:
  i) has the number 1 on it
  ii) has one of the numbers 1 to 50 on it
  iii) has one of the numbers 1 to 500 on it
  iv) has the number 501 on it.

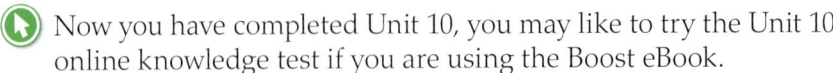

Now you have completed Unit 10, you may like to try the Unit 10 online knowledge test if you are using the Boost eBook.

Section 1 – Review

1. In the magic square on the right, the numbers 1–9 are arranged so that all the rows, columns and main diagonals add up to the same total.
   a Copy and complete the magic square by filling in the missing numbers.
   b Use your answer to (a) to complete a magic square for the numbers 11–19.
   c i) What was the total of each row, column and main diagonal in (b) above?
      ii) Explain the difference between this total and the total for each row, column and main diagonal in the original magic square.

	2	
9	5	1
	3	

2. You will need isometric dot paper for this question.
   Part of a pattern using four rhombuses is drawn on isometric dot paper below.

   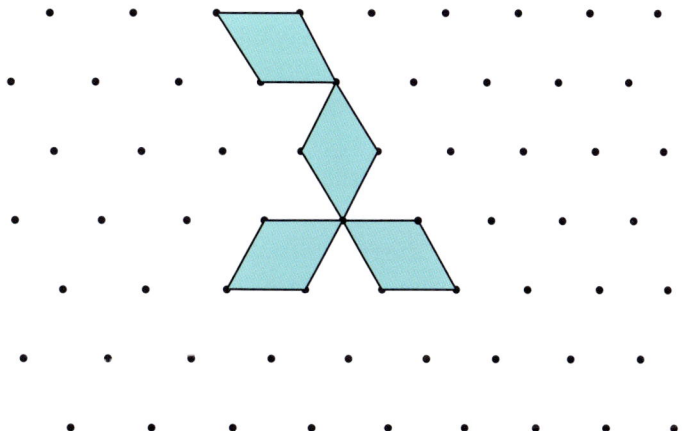

   By drawing two more rhombuses, complete the pattern so that it has a rotational symmetry of order 3.

3. a Give one advantage and one disadvantage of using a questionnaire for collecting data.
   b Give one advantage and one disadvantage of using an interview for collecting data.

4 A triangle is drawn inside a rectangle as shown.
   If the area of triangle 1 is half the area of triangle 2,
   calculate the length x.

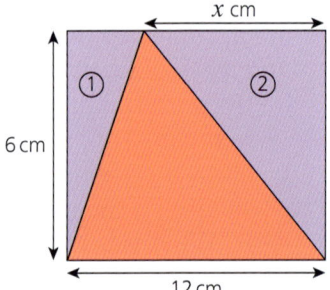

5 A student wants to work out the answer to the following
   calculation: $3+5\times4$
   He writes the following steps: $3+5\times4$
   $=8\times4$
   $=32$
   a i) Explain what mistake he has made.
      ii) What should the correct answer be?
   b By inserting any brackets necessary, rewrite the calculation so
      that the answer is 32.

6 The formula for the area (A) of a trapezium is given as
   $A=\frac{1}{2}(a+b)h$, where a, b and h are the lengths shown in
   the diagram.
   Calculate the area of the trapezium if $a=8$ cm, $b=13$ cm and
   $h=5$ cm.

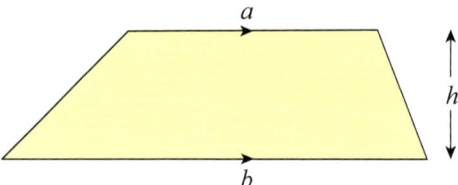

7 Eduardo collects data about the number of people in cars passing his house one afternoon. He presents the results both as a pie chart and a waffle diagram as shown.
Explain whether the diagrams are showing the same results. Justify your answer fully.

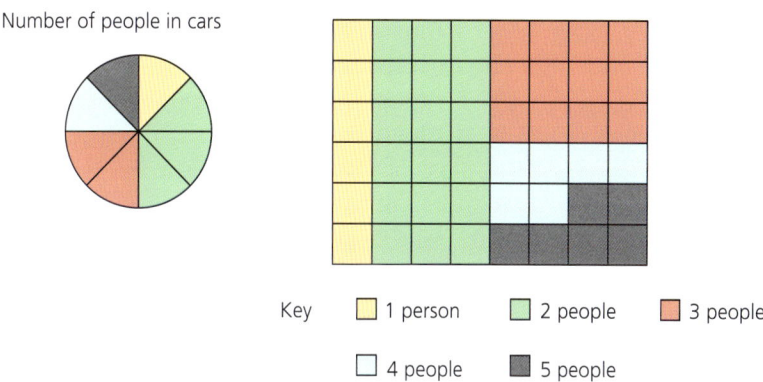

Number of people in cars

Key ☐ 1 person ☐ 2 people ☐ 3 people
     ☐ 4 people ☐ 5 people

8 Two cubes P and Q are placed on top of each other. The volume of P is eight times the volume of Q. If the combined volume of the two cubes is 243 cm³, calculate the side length of each of the two cubes.

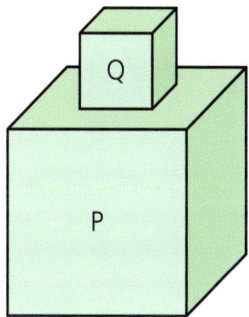

75

SECTION 1

9 A four-digit number is given below. Unfortunately one of the digits is covered with ink.

3 * 4 8

a i) If the four-digit number is divisible by 4, what digit can the missing one be?
   ii) Justify your answer.
b i) If the four-digit number is *also* divisible by 9, what number must the missing one be?
   ii) Justify your answer.

10 A class of students is asked what their favourite sport is. The results are shown in the table.

	Football	Hockey	Tennis	Basketball	Other
Boys	5	1	2	4	3
Girls	3	3	5	1	3

A student is chosen at random. What is the probability that:
a it is a girl
b it is a basketball player
c it is a boy who likes hockey
d it is a girl whose favourite sport is not tennis?

SECTION 2

History of mathematics – The development of algebra

The roots of algebra can be traced to the ancient Babylonians, who used formulae for solving problems. However, the word algebra comes from the Arabic language. Muhammad ibn Musa al-Khwarizmi (AD790–850) wrote Kitab al-Jabr (The Compendious Book on Calculation by Completion and Balancing), which established algebra as a mathematical subject.

He is regarded as the father of algebra.

11 Rounding and estimation – calculations with decimals

- Round numbers to a given number of decimal places.
- Estimate, add and subtract positive and negative numbers with the same or different number of decimal places.
- Use knowledge of place value to multiply and divide whole numbers and decimals by any positive power of 10.
- Estimate, multiply and divide decimals by whole numbers.

Decimal places

When using a calculator to do calculations, the calculator will give the answer to as many decimal places as will fit on the screen. For example, working out 355 ÷ 7 on the calculator will return the answer 50.71428571. This answer, although precise, would rarely be written down in full. Usually answers like this are rounded.

One way of rounding a number is to write it to a given number of decimal places. This refers to the number of digits written after the decimal point.

> **Worked example**
>
> The length of this model car is 7.864 cm. Write 7.864
> a to the nearest whole number
> b to one decimal place
> c to two decimal places.
>
> Draw a number line to help you.
>
> a
>
>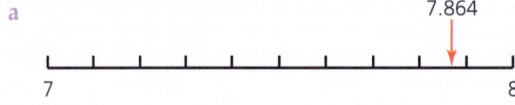
>
> 7.864 is closer to 8 than it is to 7, so 7.864 written to the nearest whole number is 8.
>
> b
>
>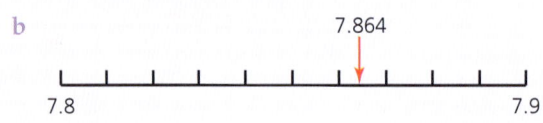
>
> 7.864 is closer to 7.9 than it is to 7.8, so 7.864 written to one decimal place is 7.9.

This number line has been divided into tenths

This number line has been divided into hundredths

KEY INFORMATION
A number written to one decimal place has one digit after the decimal point.

11 Rounding and estimation – calculations with decimals

 This number line has been divided into thousandths

c
```
        7.864
    ↓
|—|—|—|—|—|—|—|—|—|—|
7.86                7.87
```

7.864 is closer to 7.86 than it is to 7.87, so 7.864 written to two decimal places is 7.86.

Remember: To round to a certain number of decimal places, look at the next digit after the one in question. If that digit is 5 or more, round up. If it is 4 or less, round down.

Exercise 11.1

1. Round the following numbers to one decimal place.
   a 6.37
   b 4.13
   c 0.85
   d 8.672
   e 1.093
   f 0.063

2. Round the following numbers (i) to one decimal place and (ii) to two decimal places.
   a 4.383
   b 5.719
   c 5.803
   d 1.477
   e 3.999
   f 6.273

3. Round the following numbers (i) to two decimal places and (ii) to three decimal places.
   a 0.5682
   b 3.4765
   c 8.8467
   d 3.9995
   e 9.9999

4. Salma uses her calculator to work out an answer.
   Salma writes down the answer correct to two decimal places as 3.46.
   Below are four possible calculator screens. Which one could not be the answer on Salma's calculator?

   a [3.46083919]
   b [3.45500182]
   c [3.46500835]
   d [3.46499993]

5. A shop is offering a promotion on one of its products.
   Originally a pack of two items could be bought for $13.22. With the promotion, a pack of three items can be bought for the same amount.
   a How much was the cost of each item before the promotion?
   b How much is the cost of each item during the promotion?

6. In a race, six runners complete 100 m in the following times:
   10.402 s   10.395 s   10.382 s   10.374 s   10.404 s   10.406 s
   Which times are the same when given to two decimal places?

SECTION 2

7 Two blocks of wood A and B are joined as shown below.

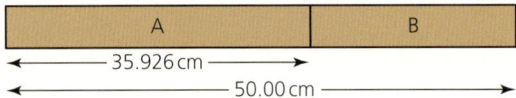

The length of block A is 35.926 cm to 3 decimal places. The total length of A and B is 50.00 cm to 2 decimal places.
If the length of block B is also given to 3 decimal places, between what range of values can it take?

Calculations and estimations with decimals

To add or subtract numbers involving decimals, we usually use a written method rather than trying to do the calculation mentally, although it is always a good idea to estimate the answer mentally as a check.

Worked examples

KEY INFORMATION
Notice how three of the numbers are written to two decimal places, whilst the other is given as a whole number. To add them it is easier to write them all to the same number of decimal places.

1 Add the following amounts of money.
$6.83 $27 $0.04 $142.30
First write the four numbers so that the decimal points line up.

```
        ¹6 .  8  3
     ¹2  7 .  0  0
         0 .  0  4
+    1   4  2 . 3  0
     1   7  6 . 1  7
```

The total of the amounts is $176.17.

The answer can be checked by estimating roughly what it should be. This can be done by rounding each number to a number which is easier to add mentally, i.e. the sum can be estimated by adding $7+30+0+140=177$.

KEY INFORMATION
If the total of any column is 10 or more, the tens are carried over to the next column (as shown in blue).

2 Subtract $72.87 from $200.

```
  ¹2  ⁹1̶0̶  ⁹1̶0̶ . ⁹1̶0̶  ¹0
        7   2  .  8   7
   1    2   7  .  1   3
```

The difference between the amounts is $127.13.
The answer can be checked by estimation using $200-70=130$.

KEY INFORMATION
If the digit on top is smaller than the digit below, a **ten** is carried over from the next column.

Exercise 11.2

1. i) Estimate the answer to the following calculations.
   ii) Work out the answer to each calculation.
   - a $27.43 + $89.29
   - b $100 − $57.57
   - c $4.62 + $0.82 + $105.62
   - d $500 − $46.30 − $3.88
   - e $26.43 + $102.11 − $37.28

2. A family of four people check in their suitcases at the airport. The weights of the four cases are 18.5 kg, 26 kg, 15.4 kg and 23.7 kg.
   - a Calculate the total weight of the four cases.
   - b The weight limit for the four cases is 100 kg. Calculate how much extra weight the family could have carried. Show all your working clearly.

3. A boy is 81.4 cm tall. He needs to be at least 1 m tall before he can go on a particular ride at a theme park. How much more does the boy need to grow before he will be allowed on the ride?

4. A bridge over a road is 3.2 m high at its lowest point. A lorry 2.65 m high passes under the bridge. Calculate the height of the gap between the bridge and the lorry, giving your answer in centimetres.

5. A pizzeria offers the following pizzas for sale.
   Margherita $6.25
   La Reine $8.15
   Veneziana $6.85
   Fiorentina $7.95
   A group of friends order one Margherita pizza, two La Reine pizzas, one Veneziana pizza and three Fiorentina pizzas.
   - a Calculate the total cost of the pizzas.
   - b They pay for the pizzas with $60. Calculate the amount of change they are due. Show all your working clearly.

Multiplying and dividing by powers of 10

10×10 can be written using powers as 10^2 and is equal to 100.

$10 \times 10 \times 10$ can be written using powers as 10^3 and is equal to 1000.

A **power of 10** occurs when 10 is multiplied by itself a number of times.

Multiplying a number by power of 10 results in its digits moving to the left. For example,

$28 \times 10 = 280$ $34.56 \times 10 = 345.6$

$28 \times 10^2 = 2800$ $34.56 \times 10^2 = 3456$

$28 \times 10^4 = 280\,000$ $34.56 \times 10^4 = 345\,600$

KEY INFORMATION
10^2 is read as 'ten to the power of two' or 'ten squared'.

KEY INFORMATION
10^3 is read as 'ten to the power of three' or 'ten cubed'.

SECTION 2

Similarly, dividing a number by a power of 10 results in its digits moving to the right. For example,

$28 \div 10 = 2.8$ $\qquad$ $34.56 \div 10 = 3.456$

$28 \div 10^2 = 0.28$ $\qquad$ $34.56 \div 10^2 = 0.3456$

$28 \div 10^5 = 0.00028$ $\qquad$ $34.56 \div 10^5 = 0.0003456$

Exercise 11.3

1. Multiply the following numbers by 10.
   a 630 b 4.6 c 0.84 d 0.065 e 1.07
2. Multiply the following numbers by 10^2.
   a 45 b 7.2 c 0.96 d 0.0485 e 6.033
3. Divide the following numbers by 10.
   a 680 b 72 c 8.9 d 0.64 e 0.054
4. Divide the following numbers by 10^2.
   a 3500 b 655 c 5.62 d 0.8 e 0.034
5. Find the value of the following.
   a 46×10^3
   b $6.4 \div 10^2$
   c 6.8×10^3
   d $46 \div 10^3$
   e 3.8×10^4
   f 0.0084×10^5
   g 0.7×10^6
   h $950 \div 10^5$
   i $0.0845 \div 10^4$
   j $4 \div 1000$
6. Each stage of this calculation chain involves a multiplication or division by a power of 10. Write down the missing powers of 10. The first has been done for you.

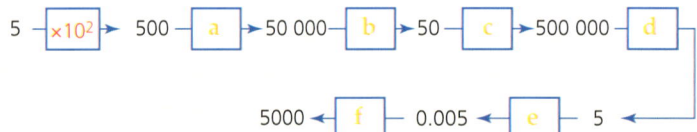

7. Starting with 0.2, four operations involving powers of 10 are used to give a final answer of 2, as shown in the calculation chain below.

   The four operations used are as follows $\quad \div 10^2 \quad \times 10^4 \quad \div 10^3 \quad \div 10^2$
   a Arrange the four operations in the calculation chain to give the correct final answer.
   b Can you arrange the four operations in a different way to give the same final answer?
   c Does the order of the operations matter? Give a **convincing** reason for your answer.
   d The result halfway is known, as shown below.

   Write a possible order for the operations now.

Multiplications, divisions and estimation with decimals

Written methods for multiplication and division involving decimals are similar to those used for whole numbers.

> **Worked example**
>
> Multiply 172×35
> - Arrange the numbers above each other, such that the units line up.
>
> ```
>         Hundreds Tens Units
>            1    7    2
>        ×        3    5
> ```
>
> - Multiply each of the digits in 172 by the 3 (which is 3 tens, i.e. 30).
>
> ```
>            1    7    2
>        ×        3    5
>       ─────────────────
>        5    1    6    0      ← The zero indicates that 172
>          2                      was multiplied by 30 rather
>                                 than by 3.
> ```
>
> - Multiply each of the digits in 172 by the 5.
>
> ```
>            1    7    2
>        ×        3    5
>       ─────────────────
>        5    1    6    0
>             ₃8   ₁6   0
> ```
>
> - Add the two answers together.
> The answer is 6020.
>
> ```
>            1    7    2
>        ×        3    5
>       ─────────────────
>        5    1    6    0
>             8    6    0
>       ─────────────────
>        6    0    2    0
>       ₁    ₁
> ```
>
> This method is known as **long multiplication**.

SECTION 2

If the question was instead to multiply 17.2×35, the method would be exactly the same. However, as 17.2 is 172÷10, the answer too must be divided by 10.

Therefore 17.2×35=602.

A good habit to do is to check your answer by estimation. In this case 17.2×35 can be approximated to 20×30=600.

Another method of doing multiplication with pen and paper is to use the **grid method**.

a 172 is a three-digit number.
35 is a two-digit number.
- Draw a 3×2 grid and draw diagonals as shown.

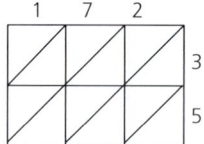

- Multiply each digit of 172 by each digit of 35 and enter each answer in the corresponding part of the grid.

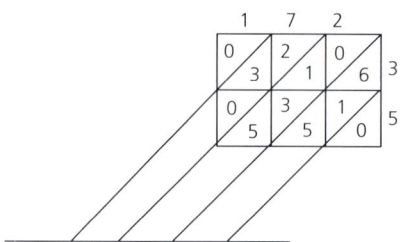

- Extend the diagonals.

- Add each of the digits in each diagonal together.

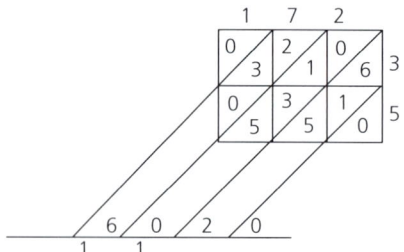

11 Rounding and estimation – calculations with decimals

The answer, as before, is 6020.
Similarly, here, as 17.2 is 172 ÷ 10, the answer must be divided by 10 also.
Therefore 17.2 × 35 = 602.

b Calculate 736 ÷ 23.
In other words, how many times does 23 go into 736?
- Set up the division as follows

$$23\overline{)736}$$

- Working from left to right, work out how many times 23 goes into each digit.
Firstly, how many times does 23 go into 7?

$$23\overline{)7^{7}36}\quad\quad 0$$

0 times therefore with 7 left over (*remainder*)

- How many times does 23 go into 73?

$$23\overline{)7^{7}3^{4}6}\quad\quad 03$$

3 times with 4 remainder

- How many times does 23 go into 46?

$$23\overline{)7^{7}3^{4}6}\quad\quad 032$$

twice with 0 remainder

Therefore 736 ÷ 23 = 32.

If the calculation had instead been to work out 73.6 ÷ 23, the method would be exactly the same. As 73.6 is 736 ÷ 10, the answer too must be divided by 10.

Therefore 73.6 ÷ 32 = 3.2.

Once again, estimating the answer is good mathematical practice.

Using knowledge of times tables, a good approximation for 73.6 ÷ 32 is instead to work out 75 ÷ 25 = 3.

> **KEY INFORMATION**
> The answer to 58.35 ÷ 7 = 8.33571 to five decimal places. Knowledge of the times tables led to the second improved estimation.

> **Worked example**
>
> Estimate the answer to 58.35 ÷ 7.
>
> The simplest estimation is to treat it as 60 ÷ 10 = 6.
>
> Knowledge of the times tables leads to a better estimation of 56 ÷ 7 = 8.

SECTION 2

Exercise 11.4

1. In each of the following multiplications:
   i) estimate the answer
   ii) multiply each pair of numbers using either long multiplication or the grid method.
   a 2.7×31
   b 4.6×21
   c 5.7×69

2. In each of the following multiplications:
   i) estimate the answer
   ii) multiply each pair of numbers using either long multiplication or the grid method.
   a 36.1×21
   b 4.06×38
   c 62×5.92
   d 0.87×193

3. Without working out the exact answers, state which of the following calculations must definitely be wrong. Justify your answers.
   a 72.4×15=1086
   b 123.2×9.3=1415.76
   c 94.35÷3=31.45
   d 127.8÷18=101.1

4. A charity holds a raffle. Each raffle ticket costs $4.50. They manage to sell 4238 raffle tickets. The first prize of the raffle is $10 000 and there are also two second prizes of $2500 each. Calculate how much profit the charity makes from raffle. Show all your working clearly.

5. In each of the following divisions:
   i) estimate the answer
   ii) work out the answer.
   a 52.5÷15
   b 96.0÷40
   c 93.5÷17
   d 97.2÷12
   e 72.6÷22
   f 91.8÷18

6. Work out the value of the letter in each of the following.
   a $31 \times x = 37.2$
   b $y \times 14 = 86.8$
   c $p \times 8 = 65.6$
   d $85.1 \div x = 37$
   e $83.3 \div a = 17$
   f $9.81 \div d = 9$

7. A carpenter has a plank of wood of length 3.52 m. He cuts the wood in to 7 pieces of equal length. Calculate the length, in cm, of each piece correct to:
   a 1 decimal place
   b 2 decimal places
   c 3 decimal places.

Now you have completed Unit 11, you may like to try the Unit 11 online knowledge test if you are using the Boost eBook.

Mode, mean, median and range

- Use knowledge of mode, median, mean and range to summarise large data sets.
- Interpret data, identifying patterns, within and between data sets to answer statistical questions.

 ## Averages

When asked if she was good at maths, a student said she was average. A boy thought he was of average height for his class. What they were saying was that, in the group being discussed, they were somewhere in the middle.

This is useful in general conversation, but may not be specific enough for all purposes. 'Average earnings', 'average speed' and 'average throw' may mean different things depending on how the data is considered.

Mathematicians have realised that there are three useful ways of looking at the average of a set of data. They are called the **mean**, the **median** and the **mode** of the data.

Mean

You will probably be familiar with this measure of average. Look at this diagram showing strips of squares.

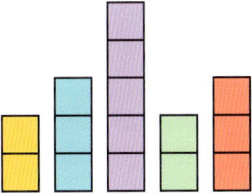

If the strips are rearranged so that there are the same number of squares in each strip, how many will there be in each? These squares have been rearranged to show this.

This problem can be solved by moving the squares around, as shown, but there is another way. We can count the total number of squares, 15, and divide by the number of strips, 5, giving a mean of 3.

SECTION 2

KEY INFORMATION
Notice that the mean does not have to be a whole number, such as in this case scoring 2.8 goals.

LET'S TALK
Is it possible to calculate the mean of discrete, continuous and categorical data?

Why? Why not?

KEY INFORMATION
If there are an even number of pieces of data, the median will fall between the middle two values.

If there is an odd number of pieces of data there will be a single middle value.

In general,

$$\text{mean} = \frac{\text{sum of all the values}}{\text{number of values}}$$

Worked example

A football team scores these numbers of goals in its matches.

1 3 2 4 4 2 2 5 2 3

Calculate the mean number of goals scored per match.

The mean number of goals scored per match is:

$$\text{mean} = \frac{\text{sum of all the goals}}{\text{number of matches}}$$

$$\text{mean} = \frac{28}{10} = 2.8 \text{ goals per match}$$

Median

Another way of looking at the average number of goals is to put the scores in order of size as shown here.

1 2 2 2 2 : 3 3 4 4 5

Since there are ten scores, the middle would be between the fifth and sixth values, which are 2 and 3. So 2.5 is the median.

Mode

A third way to look at the scores is to see which value occurred **most often**.

This is called the mode (also known as the **modal value** or **modal number**). In the example above it is 2 because the team scored 2 goals more often than any other number.

However, there are limitations to these forms of data analysis as highlighted in the next section.

12 Mode, mean, median and range

> **Worked example**
>
> Two basketball teams A and B each play seven matches against various opponents. The number of points they scored in each match are given in the table below.
>
> | Points scored | | | | | | | |
> |---|---|---|---|---|---|---|---|
> | **Team A** | 12 | 38 | 50 | 54 | 54 | 68 | 102 |
> | **Team B** | 54 | 54 | 54 | 54 | 54 | 54 | 54 |
>
> Calculate the mean, median and modal score for each team.
>
> Team A   Mean $= \frac{12+38+50+54+54+68+102}{7} = 54$
>
>   Median is   12   38   50   (54)   54   68   102
>
>   Mode is the most common number 54
>
> Team B   Mean $= \frac{54+54+54+54+54+54+54}{7} = 54$
>
>   Median is   54   54   54   (54)   54   54   54
>
>   Mode is the most common number 54
>
> The values for mean, median and mode are exactly the same for each team and yet their scores across all seven games are very different. Another form of analysis is also to consider the **range** of the data.

Range

A further value, which is often useful, is the range of the data. This gives a measure of how spread out the data is.

This table shows the times taken, in minutes, by two runners over a set distance.

Runner 1	10	10	10	12	12	12	12	12	13	13
Runner 2	8	8	9	10	12	12	12	12	15	18

A summary of the mean, median and mode for each runner gives us little information with which to distinguish between the two.

	Mean	Median	Mode
Runner 1	11.6	12	12
Runner 2	11.6	12	12

The range of each runner's results, however, does give us some more information.

SECTION 2

Range = highest value − smallest value

Range for Runner 1 = 13 − 10 = 3 minutes

Range for Runner 2 = 18 − 8 = 10 minutes

The first runner is consistent. This means his or her times are all quite close together.

The second runner is less consistent as his or her times are more widely spread.

> **LET'S TALK**
> Which runner might a coach choose in:
> a an individual event
> b a team event?
> Sometimes less consistent sportswomen and sportsmen are chosen because they are described as 'match winners'. What do you think this means?

Exercise 12.1

1. Find the mean, median, mode and range of each set of numbers.
   a 1 1 2 3 3 4 4 4 5 5
   b 3.2 4.8 5.6 5.6 7.3 8.9 9.1
   c 1 2 3 4 4 3 2 4 2 3 6 4 0

2. Two discus throwers keep a record of their best throws (in metres) in the last ten competitions.

Discus thrower A	32	34	32	33	35	35	32	36	36	35
Discus thrower B	32	30	38	38	33	34	36	38	34	32

   As a coach, you can only choose one of them for the next competition.
   Which would you choose? Justify your choice mathematically.

3. Measure the height of the students in your class.
   a Find the mean, median, mode and range of heights of the students.
   b As one of the group, are you average or not?
   c Give mathematical reasons to describe your position in the group.

4. Five cards are numbered as shown.

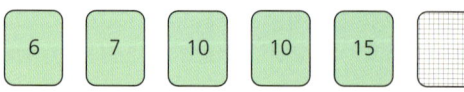

   A sixth card is covered.
   a Calculate the mean, median and mode of the five known cards.
   b If the sixth card has a number greater than 15, which values of the mean, median or mode will be affected? Justify your answer.
   c If the sixth card has a number less than or equal to 6, which values of the mean, median or mode will be affected? Justify your answer.

12 Mode, mean, median and range

5 The year of manufacture of five cars parked in a car park are as shown below.

One of the cars drives off.
If the mean year of manufacture of the remaining four cars is 1990, work out which car drove off. Justify your answer.

6 The mean mass of the 15 players in a rugby team is 85.2 kg. The mean mass of the team plus a substitute is 85.4 kg. What is the mass of the substitute?

7 After eight matches a basketball player has a mean of 27 points. After ten matches his mean was 31 points. How many points in total did he score in his last two matches?

8 A factory makes cans of tomatoes. On average, a full can should have a mass of 410 g. A sample of 20 cans is tested. The masses are shown below.

| 410 | 410 | 411 | 412 | 408 | 411 | 409 | 414 | 416 | 410 |
| 410 | 412 | 413 | 415 | 410 | 415 | 409 | 410 | 412 | 411 |

Does the canning machine need to be adjusted? Give reasons for your answer.

9 a i) Write down five possible numbers with a median of 10 and a range of 12.
   ii) Explain if a different combination of five numbers is possible.
  b i) Write down five possible numbers with a median of 10, a range of 12 and a mode of 6.
   ii) Explain if a different combination of five numbers is possible.
  c i) Write down five possible numbers with a median of 10, a range of 12, a mode of 6 and a mean of 11.
   ii) Explain if a different combination of five numbers is possible.

KEY INFORMATION
Note that not all answers are possible in this question.

10 The following set of five numbers has an interesting property in that the mean = median = mode.
11 14 14 15 16
  a Write down five numbers in which:
     i) the mode is less than the median and the median is less than the mean
     ii) the mean is less than the mode and the mode is less than the median
     iii) the mode is less than the mean and the mean is less than the median.
  b Justify why one of the above is not possible.

Calculations using frequency tables

In Unit 7 you saw that data is often presented in a frequency table. Calculations of average can still be made when large amounts of data are presented in this way.

SECTION 2

> Note that the question does not explain how the students were selected. These students may not be a random sample and therefore not representative of the school as a whole.

Worked example

a A school carries out a survey to find out how many siblings the students have.

Some students are chosen and the results are shown in the frequency table.

Number of siblings	Frequency
0	4
1	8
2	4
3	2
8	1

Calculate the mean, median, mode and range of the number of siblings for the class.

The mean number of siblings is:

$$\text{mean} = \frac{\text{total number of siblings}}{\text{total number of students}}$$

There are four students with no siblings, eight students with one sibling, four students with two siblings etc. So the total number of siblings is:

$$(4 \times 0) + (8 \times 1) + (4 \times 2) + (2 \times 3) + (1 \times 8) = 30$$

The total number of students is the sum of the frequencies, 19.

The mean = $\frac{30}{19}$ = 1.58 (2 d.p.)

Therefore, the mean number of siblings per student is 1.58.

The mode is the number of siblings with the highest frequency. So the modal number of siblings is 1.

The median *could* be calculated by arranging all 19 students in order of the number of siblings they have as shown:

0 0 0 0 1 1 1 1 1 ① 1 1 2 2 2 2 3 3 8

As there are 19 students the median value belongs to the 10th student, i.e. 1.

So the median number of siblings is also 1.

However, having to write out all the values in order is not very efficient, particularly if the data set is very large. As the numbers are already written in order in the table, the median can be calculated directly.

We know that as there are 19 students, the middle student occurs in the 10th position. As there are four students with 0 siblings and 8 students with one sibling, making a total of 12 students, the 10th student must be in the group with 1 sibling. Therefore the median is 1.

12 Mode, mean, median and range

> The range is the difference between the highest number of siblings and the smallest number of siblings.
>
> The highest number of siblings is 8.
>
> The smallest number of siblings is 0.
>
> Therefore the range is 8 − 0 = 8 siblings.
>
> b The school has a total of 475 students. Can the total number of students with one sibling be **estimated** from the sample?
>
> In this case, we are not told how the sample of 19 students used in the survey are selected.
>
> If the 19 students were randomly selected and are therefore representative of the school, then the number of students with 1 sibling can be estimated using 475 ÷ 19 = 25. Therefore the sample is $\frac{1}{25}$ th of the school population.
>
> An estimate of the number of students with 1 sibling is therefore 25 × 8 = 200.
>
> However, the sample may not have been randomly selected and if this is the case, then it is not possible to estimate the total number of students in the school with one sibling based on this sample.

KEY INFORMATION
Working out a value for the population based on a sample can only ever be an **estimate**.

Exercise 12.2

 1 The reading age of a book is based mainly on the average length of the words used in the book. The mean, median, mode and range of word length are calculated for three different sections of the book; the first paragraph, the first page and the first chapter. The results are shown below.

	Word length		
	1st paragraph	1st page	1st chapter
Mean	4.2	4.6	4.8
Median	4.5	4	4.5
Mode	3	4	4
Range	6	7	8

Which sample section is likely to give a better estimate for the average length of words in the whole book? Justify your answer.

SECTION 2

2 This frequency table shows the numbers of fish caught by the competitors in the first two hours of a fishing competition.

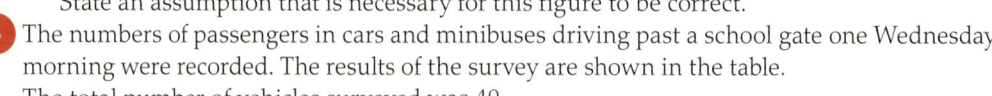

Number of fish	0	1	2	3	4	5	6
Frequency	6	20	45	70	35	10	2

a Calculate:
  i) the mean number of fish caught
  ii) the modal number of fish caught
  iii) the median number of fish caught
  iv) the range in the number of fish caught.
b A competition organiser states that if a total of 188 fish are caught in the first two hours, then 376 fish are likely to be caught in the next four hours.
State an assumption that is necessary for this figure to be correct.

3 The numbers of passengers in cars and minibuses driving past a school gate one Wednesday morning were recorded. The results of the survey are shown in the table.
The total number of vehicles surveyed was 40.

Number of passengers	Frequency
0	6
1	4
2	8
3	10
4	x
y	2

a Calculate the frequency of the vehicles with four passengers.
b The mean number of passengers carried in the vehicles was 2.75. Calculate the number of passengers represented by y in the table.
c Deduce the modal number of passengers carried.
d Calculate the median number of passengers carried.
e Are the results likely to be similar on a Saturday? Justify your answer.

Now you have completed Unit 12, you may like to try the Unit 12 online knowledge test if you are using the Boost eBook.

13 Transformations of two-dimensional shapes

- Identify reflective symmetry and order of rotational symmetry of 2D shapes and patterns.
- Reflect 2D shapes on coordinate grids in a given mirror line.
- Rotate shapes 90° and 180° around a centre of rotation.
- Use positive integer scale factors to perform and identify enlargements.

Reflective and rotational symmetry

A shape has **reflection symmetry** if it looks the same on both sides of a mirror line.

For example, this L-shape has one line of reflection symmetry.

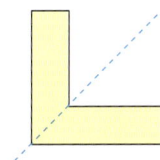

This irregular hexagon has two lines of reflection symmetry.

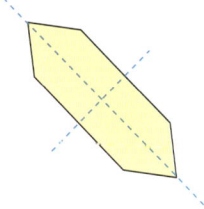

A shape has **rotational symmetry** if, during one complete revolution of 360° about the **centre of rotation**, it looks the same as the shape in its original position.

For example, this equilateral triangle has rotational symmetry of order 3, as in one complete rotation it looks the same **three** times.

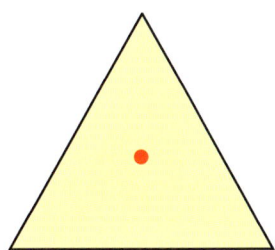

SECTION 2

LET'S TALK

Is it possible for a shape to have rotational symmetry but not reflective symmetry?

Is it possible for a shape to have reflective symmetry but not rotational symmetry?

All shapes have rotational symmetry of at least order 1, because any shape will look the same after a full rotation of 360°. So a shape which only has rotational symmetry of order 1 is considered **not** to have rotational symmetry.

This isosceles trapezium looks the same only once in a full rotation, so it is not classed as having rotational symmetry.

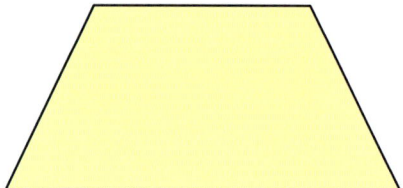

KEY INFORMATION

An isosceles trapezium has one pair of parallel sides and two pairs of equal angles.

Exercise 13.1

1. Copy and complete this **classification** table of the symmetry properties of different types of shapes.

Shape		Number of lines of reflective symmetry	Order of rotational symmetry
Square			
Rectangle			
Rhombus			
Parallelogram			
Isosceles trapezium			
Kite			
Equilateral triangle			

13 Transformations of two-dimensional shapes

Shape		Number of lines of reflective symmetry	Order of rotational symmetry
Isosceles triangle	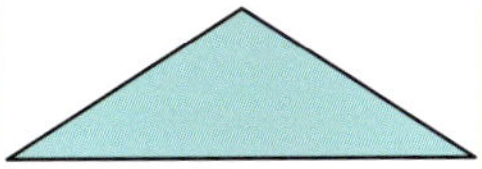		
Regular pentagon	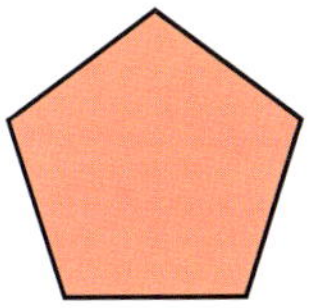		
Regular hexagon	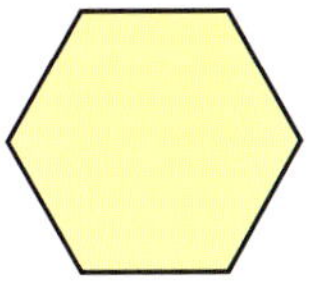		
Circle			

2 For each of these capital letters, write down
  i) the number of lines of reflection symmetry
  ii) the order of rotational symmetry.

a A

b B

c E

d H

e I

f N

g S

h Z

SECTION 2

 3 Copy each of these diagrams onto squared paper.
Draw in additional lines so that the final shape has the symmetry stated.

a
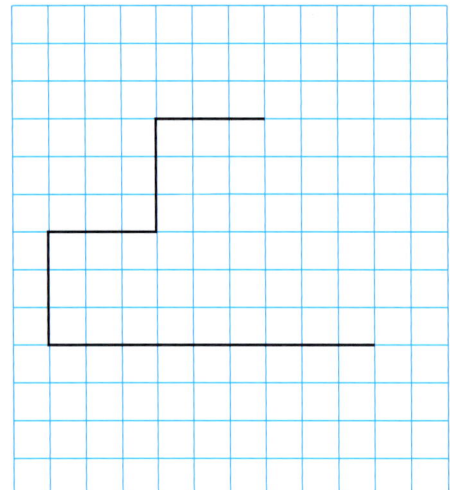

1 line of reflection symmetry.
Rotational symmetry of order 1.

c
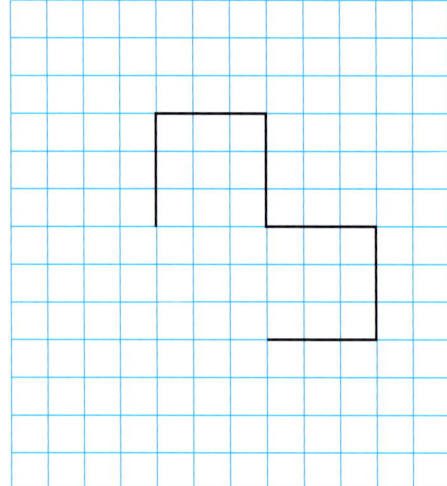

4 lines of reflection symmetry.
Rotational symmetry of order 4.

b
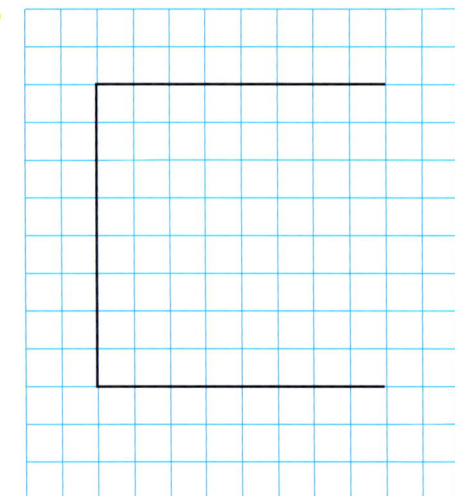

4 lines of reflection symmetry.
Rotational symmetry of order 4.

d
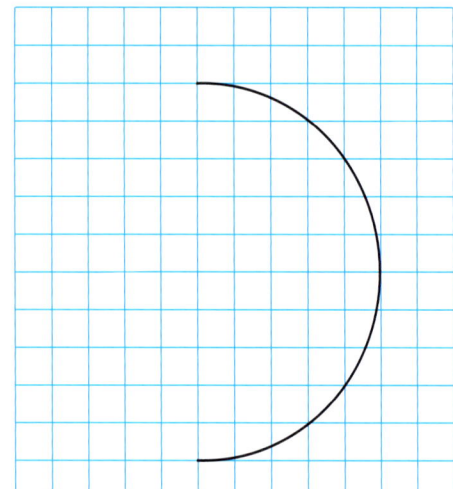

1 line of reflection symmetry.
Rotational symmetry of order 1.

13 Transformations of two-dimensional shapes

e

f

0 lines of reflection symmetry.
Rotational symmetry of order 2.

2 lines of reflection symmetry.
Rotational symmetry of order 2.

4 Make four copies of this 4 × 4 grid.

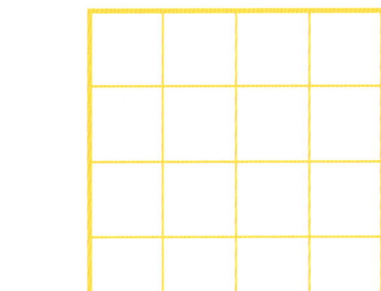

Shade in squares so that the final shape has:
a   4 lines of reflection symmetry and rotational symmetry of order 4
b   0 lines of reflection symmetry and rotational symmetry of order 4
c   2 lines of reflection symmetry and rotational symmetry of order 2
d   0 lines of reflection symmetry and rotational symmetry of order 2.

5 Three shapes formed of squares are shown below:

Arrange the three pieces together to form a shape with:
a   four lines of reflective symmetry and a rotational symmetry of order 4
b   rotational symmetry of order 2 and no lines of reflective symmetry.

SECTION 2

> **KEY INFORMATION**
> Shapes which are congruent are exactly the same size and shape. They can be reflections and/or rotations of each other. Their interior angles are equal.

Reflections on a coordinate axis

Consider the coordinate axis below with the triangle ABC and a mirror line as shown.

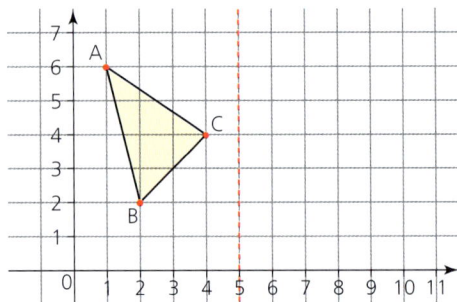

If the triangle is reflected in the mirror line it produces another triangle A'B'C' which is **congruent** to the first.

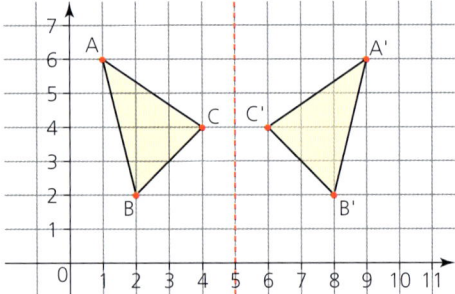

Depending on the position of the mirror line, the reflected shape (the **image**) will appear in different positions on the coordinate axis. But the image will always be congruent to the original shape (the **object**) as shown below.

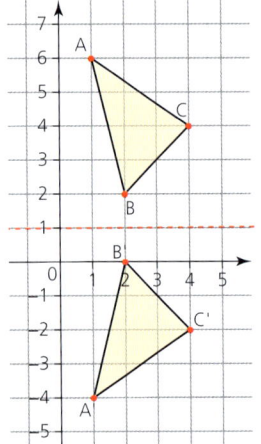 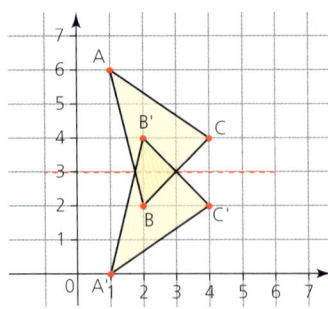 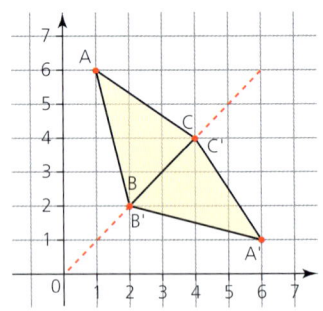

13 Transformations of two-dimensional shapes

Exercise 13.2

1 Copy the following diagram and reflect the quadrilateral in the mirror line shown.

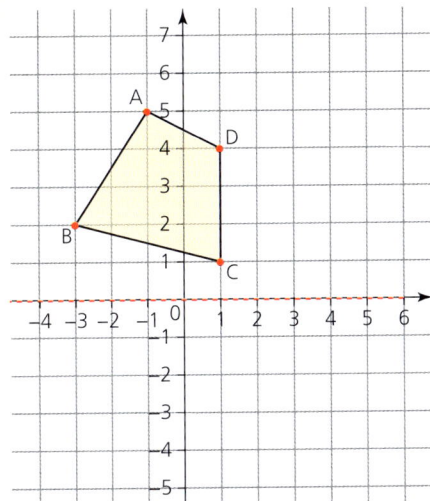

2 Copy the following diagram and reflect the shape in the mirror line shown.

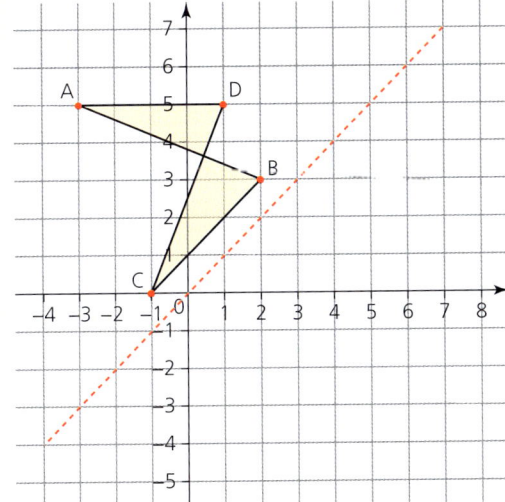

SECTION 2

3 Copy the following diagram.

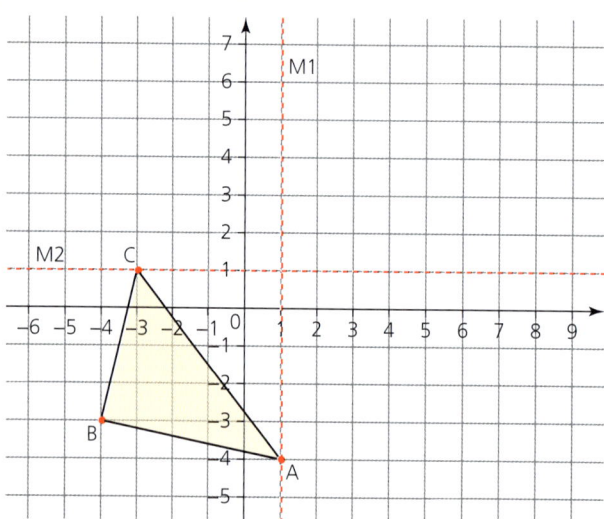

a Reflect triangle ABC in the mirror line M1 and label it A'B'C'.
b Reflect triangle A'B'C' in the mirror line M2 and label it A"B"C".
c In (a) and (b) above, the triangle ABC was first reflected in the mirror line M1 and then the mirror line M2 to map on to A"B"C". Does the order of the reflections matter to transform ABC on to A"B"C"?

4 The diagram below shows a polygon ABCD and its position A'''B'''C'''D''' after three reflections. Three mirror lines M1, M2 and M3 are shown.

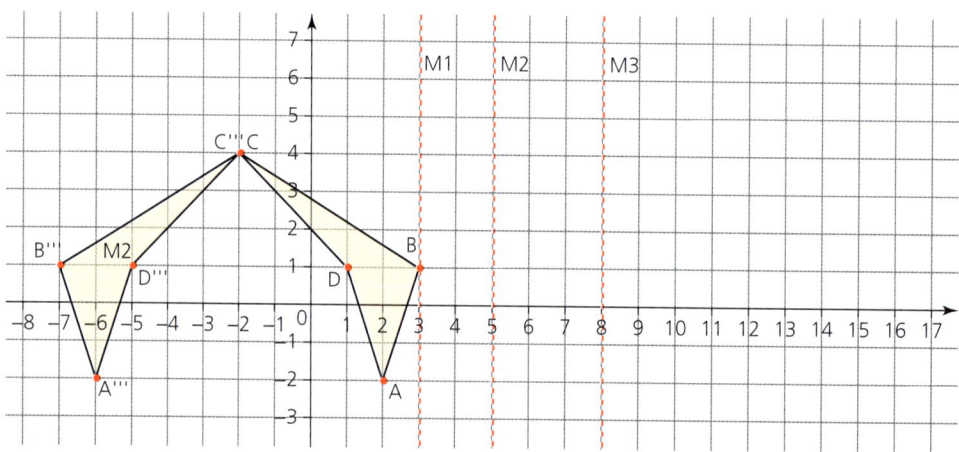

Work out which order the reflections were done in order to map ABCD on to A'''B'''C'''D'''. (Note not all of the mirror lines need to be used and some can be used more than once.)

13 Transformations of two-dimensional shapes

> **KEY INFORMATION**
> Rotations only change the position of the object, not its size or shape. Therefore rotations produce congruent shapes.

Rotation about a point

Earlier in this unit we looked at shapes with rotational symmetry. However it is also possible to rotate shapes about a point known as the **centre of rotation**.

In the diagrams below, an object A is rotated by different angles about the centre of rotation O, to produce the image B. In each case shapes A and B are congruent.

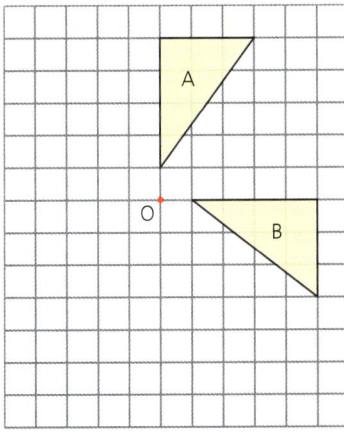
Rotation 90° clockwise

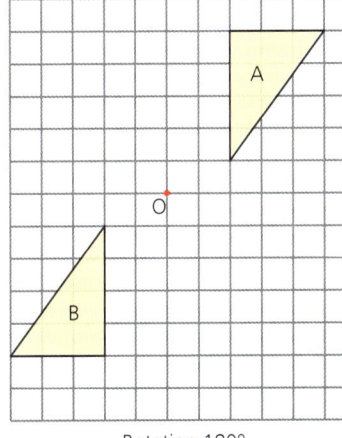
Rotation 180°

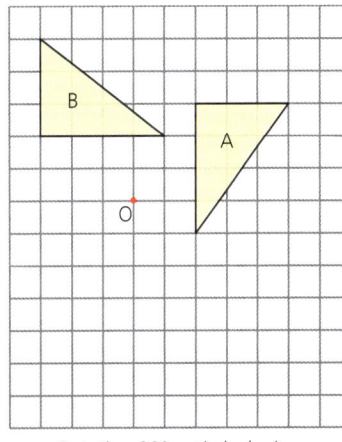
Rotation 90° anti-clockwise

> **LET'S TALK**
> Why does a rotation of 180° not need to specify whether it's clockwise or anti-clockwise?

As a way of checking, the angle between each point on the object and its corresponding point on the image must be the same about the centre of rotation.

The diagram below shows this for the 90° anti-clockwise rotation.

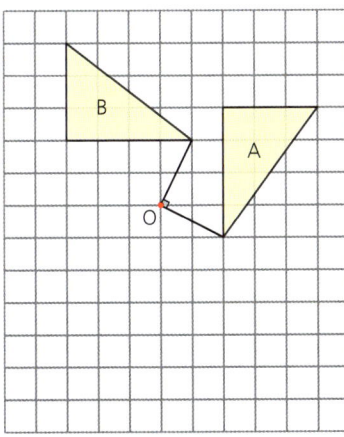
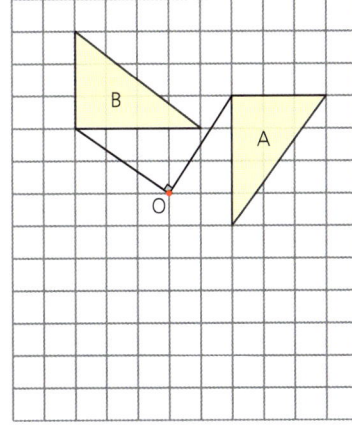
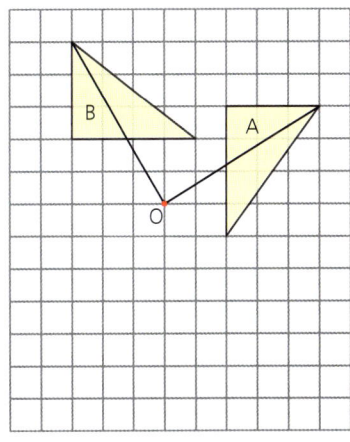

SECTION 2

Exercise 13.3

In each of the following questions, copy the grid and the object.
    a Rotate the object by the angle and direction stated about the centre of rotation O.
    b Label your image in each case P.

Note: A piece of tracing paper may help.

1

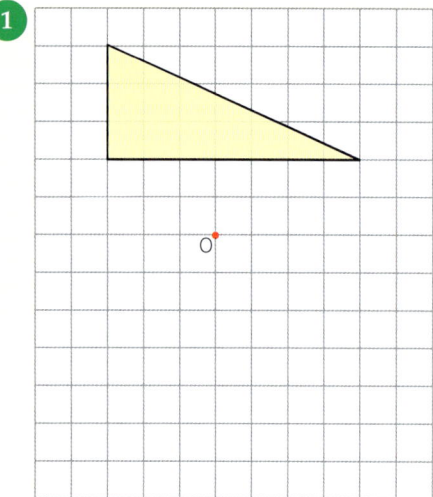

Rotation 180° about O

2

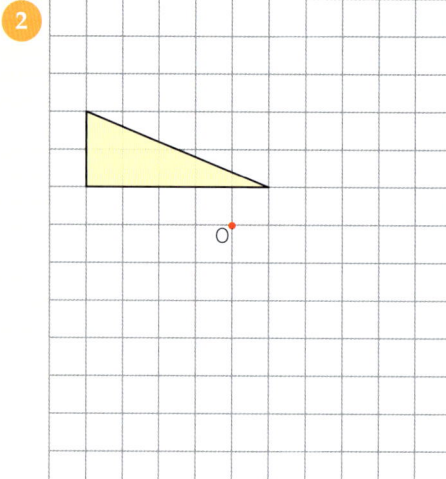

Rotation 90° clockwise about O

3

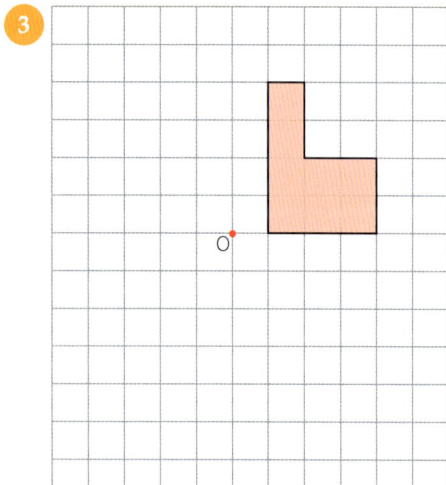

Rotation 90° clockwise about O

4

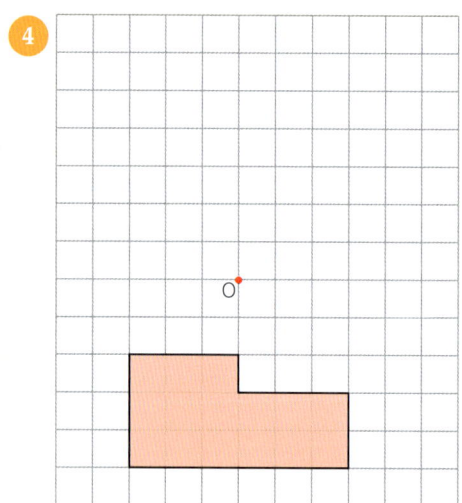

Rotation 90° anti-clockwise about O

5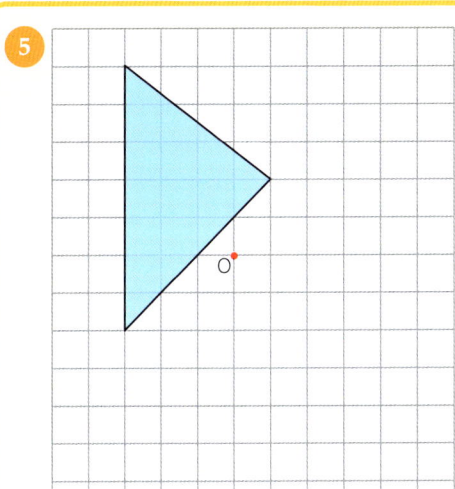

Rotation 180° about O

6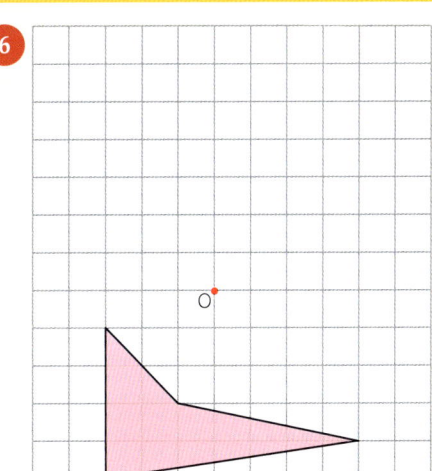

Rotation 90° anti-clockwise about O

 7 Ibrahim and Isabel are discussing the diagram below.

Ibrahim says that the square P has been rotated 180° about the origin to map on to square Q.
Isabel says that square P has been reflected in a diagonal mirror line like this ∕ passing through the origin from left to right.

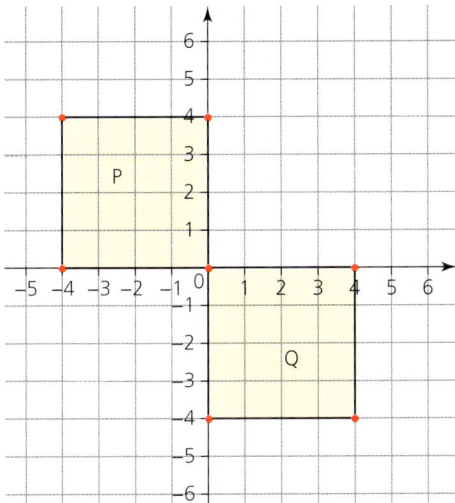

State whether you think their statements are true or if additional information is needed before you can decide.

Enlargement

In this unit so far you have studied reflection and rotation. You will also already be familiar with **translation**. These are all types of **transformation** and in each of these cases the final image is the same shape and size as the original object. All that changed was the position and/or the orientation of the image compared with the original object.

Another type of transformation is **enlargement**. With enlargement, the final image usually has a different position and size to the original object.

However, simply stating that the size changes does not give enough detail. In the examples below, picture A is the original object. Picture B shows the object stretched horizontally and picture C shows it stretched vertically.

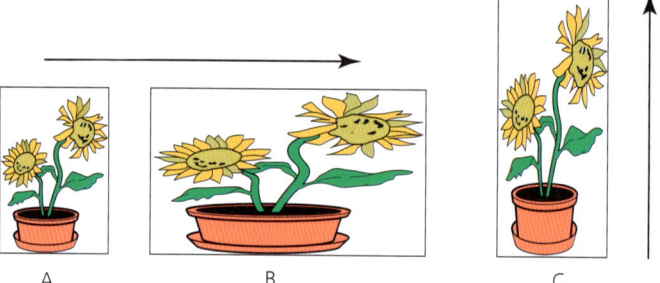

Stretches though are *not* considered enlargements. For an object to be enlarged, its lengths must all be multiplied by the same amount.

For example,

Here, both the horizontal and vertical lengths have been multiplied by 2. We say that the object has been enlarged by a **scale factor** of 2.

In an enlargement, the number which multiplies the lengths is known as the **scale factor of enlargement**.

13 Transformations of two-dimensional shapes

Worked example

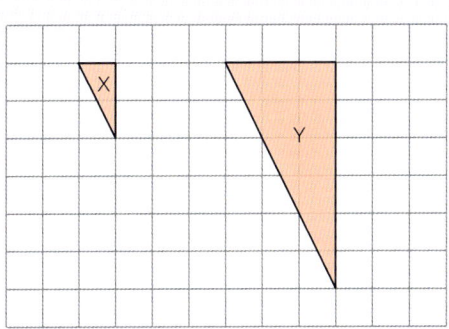

KEY INFORMATION
When a shape is enlarged all lengths are multiplied by the same factor of enlargement.

However the angles do not change.

Triangle Y is an enlargement of triangle X. Calculate the scale factor of enlargement.

Each of the lengths of triangle Y are three times that of triangle X. Therefore the scale factor of enlargement is 3.

Exercise 13.4

In each of the diagrams below, compare shape B with shape A.
a Decide whether B is an enlargement of A. If not explain why not.
b If B is an enlargement of A, calculate the scale factor of enlargement.

KEY INFORMATION
Enlarged shapes can overlap, partially or wholly.

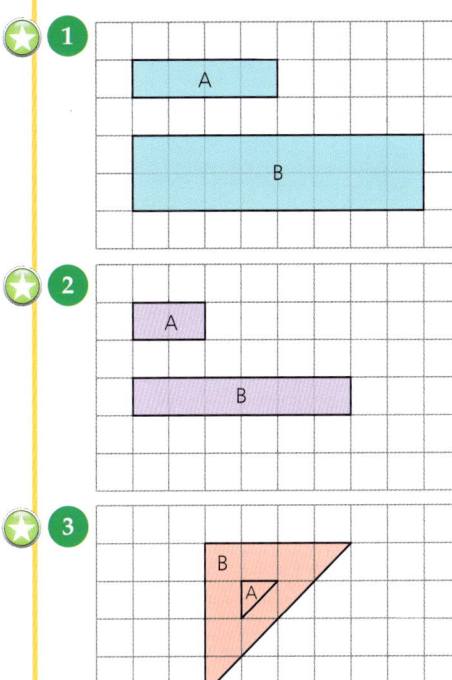

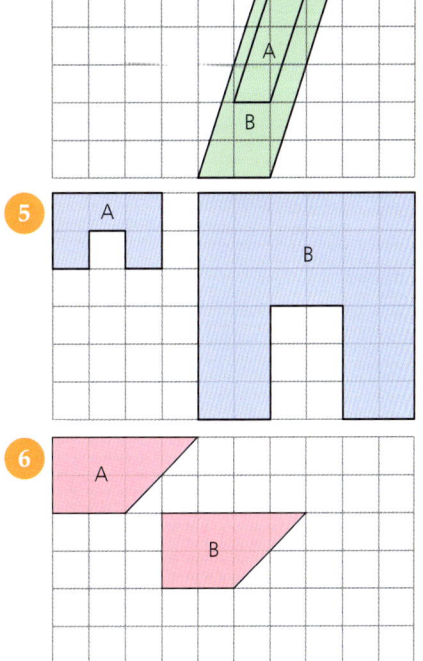

107

SECTION 2

Exercise 13.5

Copy each of the diagrams below and enlarge each of the objects by the given scale factor of enlargement.

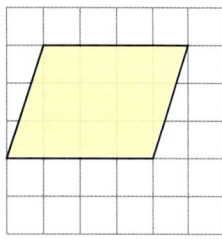

Scale factor of enlargement of 2

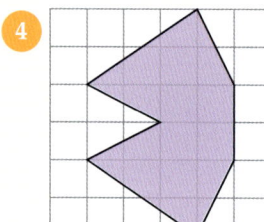

Scale factor of enlargement of 3

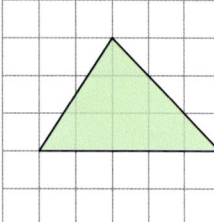

Scale factor of enlargement of 4

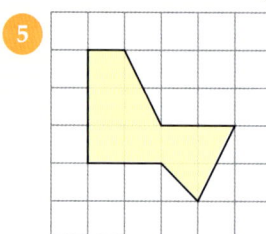

Scale factor of enlargement of 5

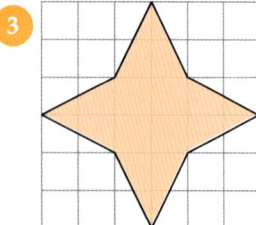

Scale factor of enlargement of 2

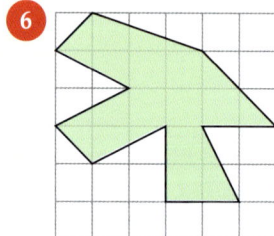

Scale factor of enlargement of 2

 7 A quadrilateral PQRS is drawn on a coordinate axes. The coordinates of each of the vertices are as follows:
P(3, 8)   Q(2, 4)   R(7, 2)   S(7, 6)
The shape is then enlarged by a scale factor of 2 and mapped on to P'Q'R'S'.
The coordinates of the vertices of P'Q'R'S' are written as follows:
P'(5, 14)   Q'(3, 6)   R'(12, 2)   S'(13, 10)
   a One of the coordinates of the vertices of P'Q'R'S' is incorrect. Which one? Justify your answer.
   b Write down the correct coordinate of this vertex.

 Now you have completed Unit 13, you may like to try the Unit 13 online knowledge test if you are using the Boost eBook.

14 Manipulating algebraic expressions

- Understand how to manipulate algebraic expressions.

Simplifying algebraic expressions

You will remember from Unit 6 that an expression is used to represent a value in algebraic form. For example,

The length of the line is given by the expression $x+3$.

There are many occasions when an algebraic expression can be simplified. It is good mathematical practice to leave expressions written in their simplest form.

Worked examples

1 The diagram below shows three containers, each with different numbers of red, blue and yellow sweets.

a A blue sweet costs b cents, a red sweet costs r cents and a yellow sweet costs y cents.

i) Write the cost of the sweets in each tray as an algebraic expression.

Tray 1: $3b + 6r + 3y$
Tray 2: $4b + 3r + y$
Tray 3: $5b + 4r + 4y$

> Remember $3b$ means $3 \times b$

ii) Write an expression for the total cost of the sweets in the three containers.

$3b + 6r + 3y + 4b + 3r + y + 5b + 4r + 4y$

This can be simplified to give $12b + 13r + 8y$

KEY INFORMATION
$12b + 13r + 8y$ is an expression with three terms. It cannot be simplified further.

2 Two rectangles are shown below:

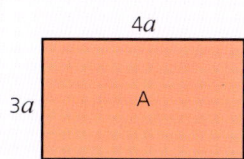
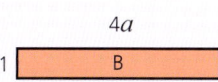

109

SECTION 2

a Write an expression for the area of each rectangle.
   Rectangle A: $4a \times 3a = 12a^2$
   Rectangle B: $4a \times 1 = 4a$

b Write an expression for the total area of the two rectangles.
   $12a^2 + 4a$

> **KEY INFORMATION**
> $12a^2 + 4a$ is an expression with two terms. a^2 and a are not **like terms** so cannot be combined. The expression cannot be simplified further.

3 Six cards are shown below, each with a different algebraic expression.

$3x^2$ $2y$ x^2 $\dfrac{y}{3}$ $\dfrac{x^2}{2}$ x

a Arrange the cards into groups of like terms.

$3x^2$ x^2 $\dfrac{x^2}{2}$ $\dfrac{y}{3}$ $2y$ x

b Simplify each group.

$3x^2 + x^2 + \dfrac{x^2}{2}$
$= \dfrac{6x^2}{2} + \dfrac{2x^2}{2} + \dfrac{x^2}{2}$
$= \dfrac{9x^2}{2}$

$\dfrac{y}{3} + 2y = \dfrac{y}{3} + \dfrac{6y}{3} = \dfrac{7y}{3}$

When adding fractions the denominators must be the same

c Write a simplified expression for the sum of all six cards.
   $\dfrac{9x^2}{2} + \dfrac{7y}{3} + x$

This could also be written as $\dfrac{9}{2}x^2 + \dfrac{7}{3}y + x$

Exercise 14.1

1 Simplify the following expressions.
   a $2a + 3a + 5a$
   b $4b - 2b + 7b$
   c $3c + 8 - 5c + 7 + 8c$
   d $4d - 5 + 8 - 2d$
   e $3e + 8e + 7 - e$
   f $6f - 7f - 3 + 4f - 8$
   g $4g + 6 - 8g - 9$
   h $3h - 7 - 9h + 11$
   i $6i - 3i - 14 + 9i - 4$
   j $3j - 4 + 8j - 11 - j$

14 Manipulating algebraic expressions

2) For each of the shapes below write a simplified expression for its perimeter.
   a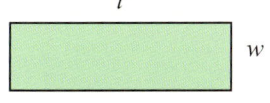

   b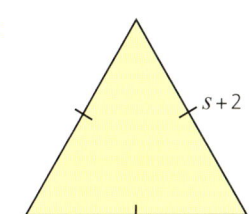

   c

   d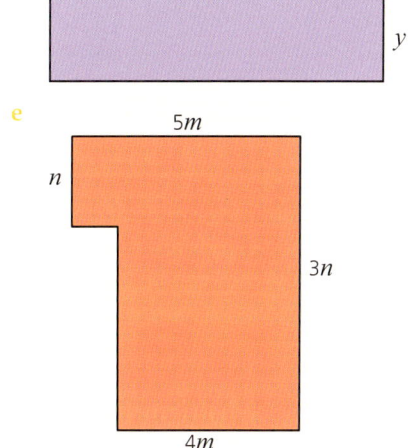

   e

3) The following rectangle has dimensions as shown.

   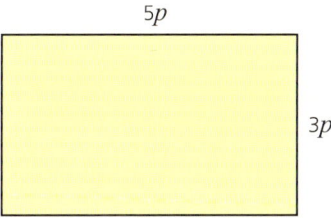

   a Write a simplified expression for the perimeter of the rectangle.
   b A small square of side length p is removed from one corner of the rectangle as shown below.

   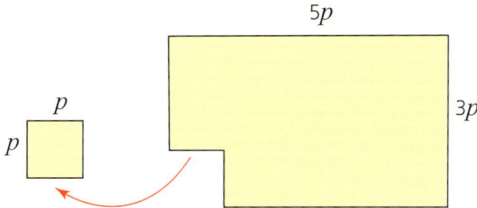

   i) What is the perimeter of the small square?
   ii) Choose one of the statements below that describes the perimeter of the remaining shape when compared with the perimeter of the original rectangle. Justify your answer using algebra.
   - The perimeter is now $2p$ less than it was before.
   - The perimeter is now $4p$ less than it was before.
   - The perimeter is unchanged.

SECTION 2

4. A child's construction set has a number of straight pieces of various lengths as shown.

$x+2$

$x+5$

$x+2$

$2x+1$

$x+3$

$2x+2$

$x+3$

$2x+4$

LET'S TALK
If x is a positive number is $2x+1$ always bigger than $x+5$?

If x can be any number is $2x+2$ always bigger than $x+2$?

Seven of the eight rods can be arranged to form an equilateral triangle.
 a i) Which seven rods are they?
    ii) Show how they can be arranged to make the triangle.
 b Write an expression for the length of each side of the triangle.

KEY INFORMATION
The three sides of an equilateral triangle are all the same length.

Multiplying simple expressions

Consider the diagram below. The rectangle has dimensions $3x+4$ and 6 units as shown.

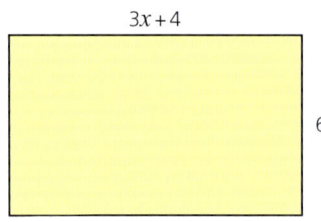

An expression for the perimeter of the rectangle is $6x+20$.

The area of the rectangle is worked out by multiplying its length by its width, therefore the expression for the area is $6(3x+4)$.

To **expand** the expression means to write it without brackets.

How to expand the expression is explained below.

The side of length $3x+4$ can be split in two with one part of length $3x$ and the other of $+4$.

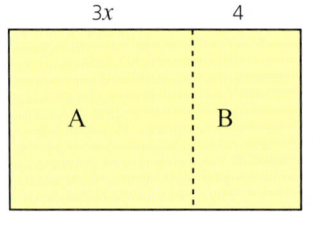

The rectangle is now split into two smaller ones A and B.

The area of rectangle A is $6 \times 3x = 18x$.

The area of rectangle B is $4 \times 6 = 24$.

Therefore the total area of the rectangle is $18x + 24$.

But the area of the rectangle was also given by the expression $6(3x+4)$, therefore $6(3x+4) = 18x + 24$

By looking at both forms you can see that the 6 outside the brackets has multiplied both of the terms inside the bracket.

14 Manipulating algebraic expressions

Worked example

Expand the following brackets.

a $4(5a - 2)$
   $20a - 8$

b $8(3a - 2b + 6)$
   $24a - 16b + 48$

Exercise 14.2

1. Write an expression for the area of the following rectangles:

   a $a + 3$, 4

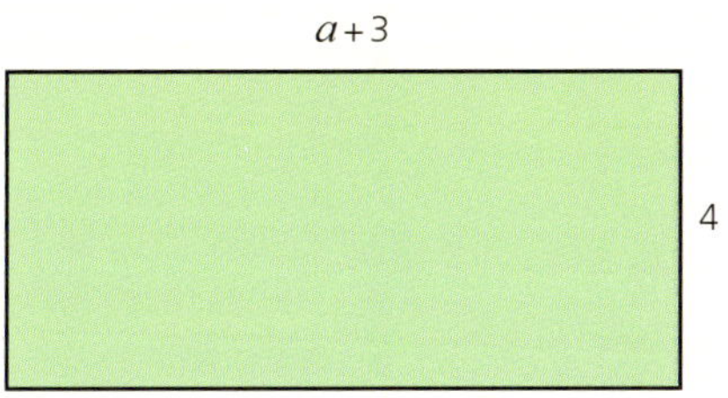

   b $b - 2$, 3

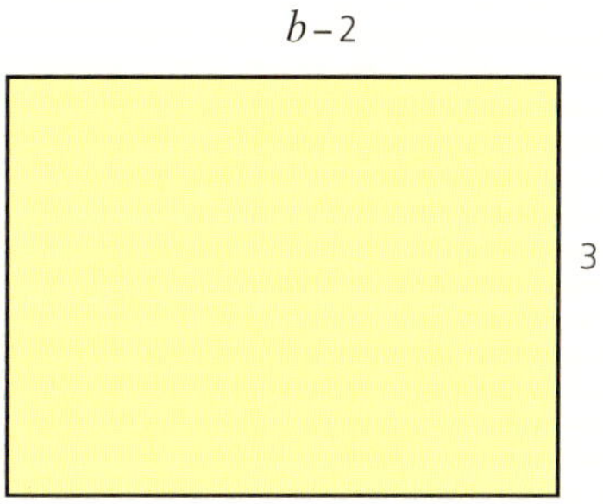

   c $3c - 1$, 5

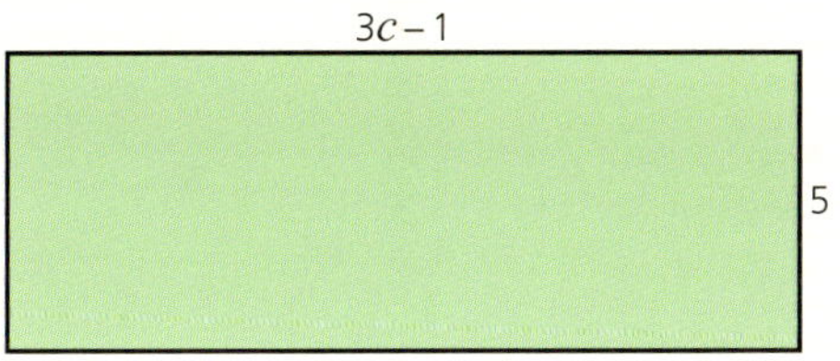

   d $\frac{d}{2} - 3$, 4

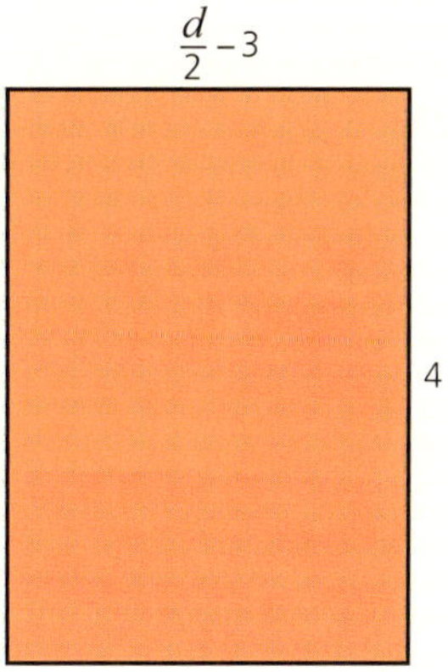

SECTION 2

2 The following right-angled triangle has an area of $5h+20$ units². Write down the length of the side marked x.

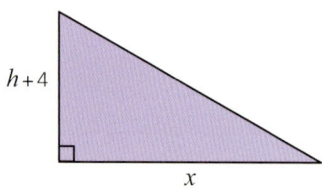

3 Write an expression for the area of the following composite shapes.

a

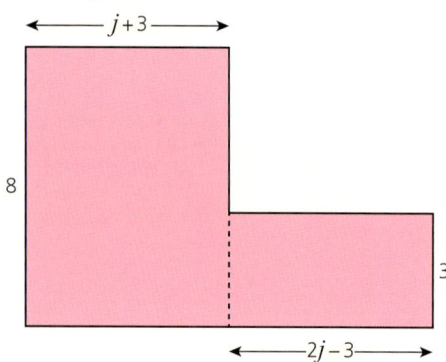

b

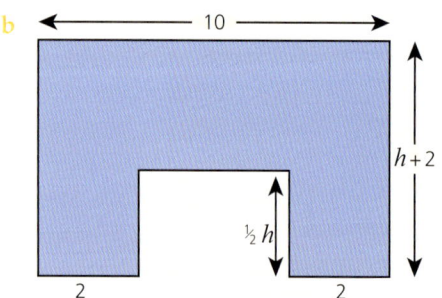

4 The following rectangle has a perimeter of $8x + 16$ units.

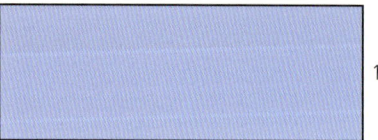

Write an expression for the area of the rectangle. Show your working clearly.

5 The composite C-shape below has dimensions as shown:

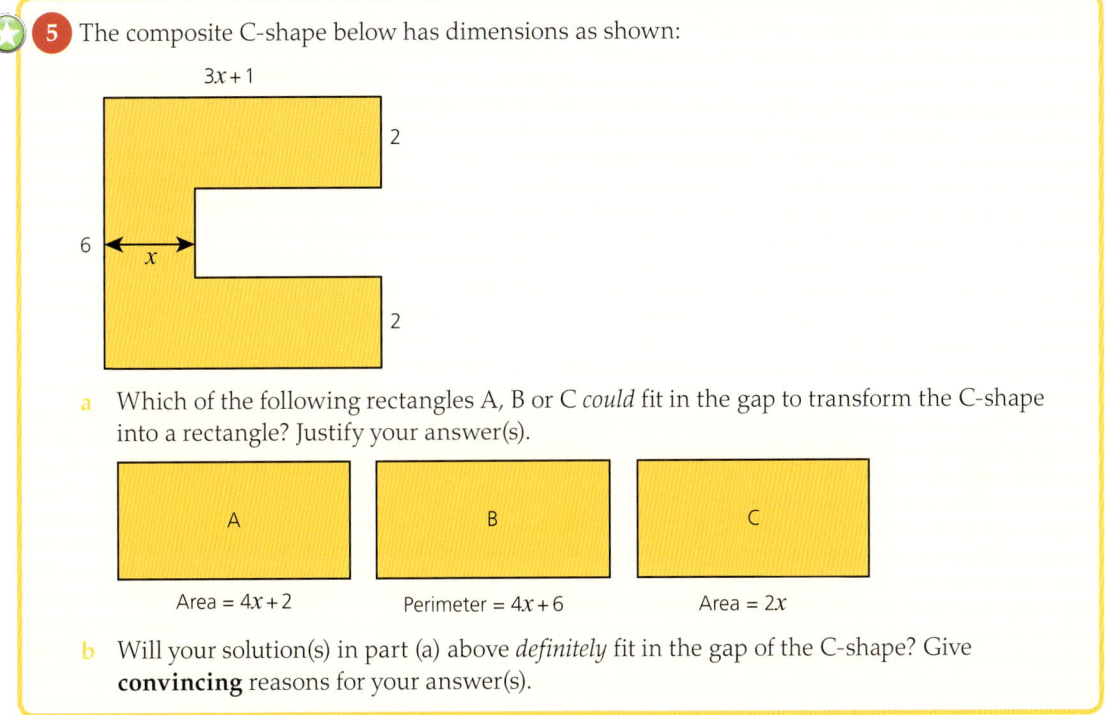

a Which of the following rectangles A, B or C *could* fit in the gap to transform the C-shape into a rectangle? Justify your answer(s).

b Will your solution(s) in part (a) above *definitely* fit in the gap of the C-shape? Give **convincing** reasons for your answer(s).

Now you have completed Unit 14, you may like to try the Unit 14 online knowledge test if you are using the Boost eBook.

Fractions, decimals and percentages

- Recognise that fractions, terminating decimals and percentages have equivalent values.
- Estimate and add mixed numbers, and write the answer as a mixed number in its simplest form.
- Estimate, multiply and divide proper fractions.
- Use knowledge of common factors, laws of arithmetic and order of operations to simplify calculations containing decimals or fractions.
- Understand the relative size of quantities to compare and order decimals and fractions, using the symbols =, ≠, > and <.

KEY INFORMATION
A terminating decimal is a decimal which ends after a certain number of decimal places.

For example, 0.2, 3.625, 1.0032

LET'S TALK
A recurring decimal is a decimal which repeats itself and therefore never ends.

Can you think of a fraction that is equivalent to a recurring decimal?

The equivalence of fractions, decimals and percentages

You will already know that some fractions can be written as decimals or as percentages,

e.g. $\frac{1}{2} = 0.5 = 50\%$ or $\frac{1}{4} = 0.25 = 25\%$

All **terminating decimals** can be written as fractions. In order to convert a terminating decimal to a fraction an understanding of place value is necessary.

Worked examples

1. Write the decimal 0.7 as a fraction in its simplest form.
   By entering 0.7 into a place value table we get:

Units	Tenths
0	7

   The '7' is worth seven-tenths. As a fraction this can therefore be written as $\frac{7}{10}$.

2. Write the decimal 0.625 as a fraction in its simplest form.
   By entering 0.625 into a place value table we get

Units	Tenths	Hundredths	Thousandths
0	6	2	5

   six-tenths, two-hundredths and five-thousandths is equivalent to 625-thousandths. As a fraction this can therefore be written as $\frac{625}{1000}$. So, $\frac{625}{1000}$ can be simplified to $\frac{5}{8}$, by dividing both the numerator and denominator by the highest common factor of 125.

Converting a fraction back to a decimal

Converting a fraction back to a decimal is relatively straightforward if the denominator is a power of 10 (i.e. 10, 100, 1000 etc.). If the denominator is not a power of 10 then it may be necessary to consider its equivalence to other fractions, where the denominator is a power of 10, or to carry out a division.

Worked examples

1. Convert $\frac{19}{100}$ to a decimal.

   As 19 is being divided by 100, $\frac{19}{100}$ can be written as 0.19.

2. Convert $\frac{8}{25}$ to a decimal.

   As 25 is a factor of 100, by multiplying both the numerator and denominator by 4, the fraction $\frac{8}{25}$ can be written in an equivalent form as $\frac{32}{100}$.

   Therefore $\frac{8}{25} = \frac{32}{100} = 0.32$.

3. Convert $\frac{50}{16}$ to a decimal.

   16 is not a power of 10, and not easily multiplied by a number to become a power of 10, therefore the fraction can be converted to a decimal by division.

   $$\begin{array}{r} 0\,3.1\,2\,5 \\ 16\,\overline{\smash{)}5\,0.0\,0\,0} \end{array}$$

   Therefore $\frac{50}{16}$ as a decimal is 3.125.

LET'S TALK

What is the smallest power of 10 that 16 is a factor of?

Discuss how the answer of 3.125 is arrived at using this method.

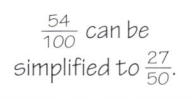
$\frac{54}{100}$ can be simplified to $\frac{27}{50}$.

Converting a percentage to a fraction or decimal

Converting a percentage to either a fraction or decimal is straightforward as it is a number expressed out of 100.

For example, 54% can be written as a fraction out of 100 as $\frac{54}{100}$ and therefore as a decimal as 0.54.

Worked example

Convert the percentage 134.5% to both a fraction and decimal.

134.5% as a fraction can be written as $\frac{134.5}{100}$. However, it is not usual to write decimals within a fraction. To eliminate the decimal, both numerator and denominator can be doubled, i.e. $\frac{134.5}{100} = \frac{269}{200}$.

As a decimal 134.5% can be written as 1.345.

SECTION 2

Exercise 15.1

1 Complete the following table showing the **equivalent fractions**, decimals and percentages.

Fraction	Decimal	Percentage
$\frac{1}{2}$		
$\frac{1}{4}$		
	0.2	
		10%
	0.375	
$\frac{21}{50}$		
		17.5%
	0.64	
		13.2%

2 The following cards show fractions, decimals and percentages.

| 0.92 | $\frac{23}{25}$ | 160% | 1.85 | $\frac{7}{8}$ | 87.5% | $\frac{8}{5}$ |

| $\frac{17}{20}$ | 185% | 0.875 | $\frac{37}{20}$ | 92% | 1.6 |

   a Group together equivalent fractions, decimals and percentages.
   b One card has no equivalent value on the other cards.
      i) Which card is this?
      ii) Write down the missing equivalent values to this card.

3 Convert each of the following percentages to a fraction.
   a 152.5%
   b 30.25%
   c 205.2%
   d 10.05%

4 A fraction in its simplest form and its equivalent decimal are shown below. However, most of the numbers are covered.

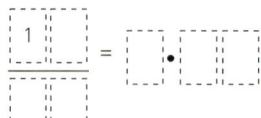

> The hidden digits, in no particular order, are as follows:
>
> $$4 \quad 5 \quad 6 \quad 6 \quad 0 \quad 2$$
>
> a   Explain why the number before the decimal point **must** be the zero.
> b   Explain why both digits in the units column of the numerator and denominator cannot be even numbers.
> c   Work out the position of all the digits.

Addition and subtraction of fractions

Recap

To add or subtract fractions with the same denominator is relatively straightforward.

For example, $\quad \frac{1}{8} \quad + \quad \frac{3}{8} \quad = \quad \frac{4}{8} \quad = \quad \frac{1}{2}$

Visually:

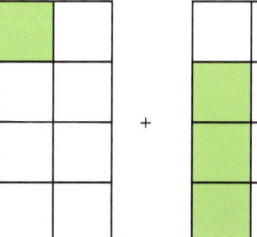  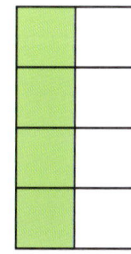 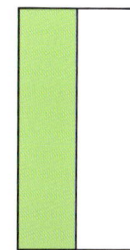

That is simply add the numerators together and keep the denominator as it is.

However adding or subtracting fractions with different denominators is a little less straightforward.

For example, $\quad \frac{1}{4} \quad + \quad \frac{2}{5}$

In order to do this, both fractions need to be converted into equivalent fractions with a common denominator. The **lowest common multiple** of both denominators is 20 (i.e. 20 is the smallest number that both 4 and 5 go in to). Therefore, we find equivalent fractions to those given, with 20 as a denominator.

$$\frac{1}{4} = \frac{5}{20} \quad \text{and} \quad \frac{2}{5} = \frac{8}{20}$$

Therefore $\frac{1}{4} + \frac{2}{5}$ is the same as $\frac{5}{20} + \frac{8}{20}$.

SECTION 2

Visually this can be shown as follows:

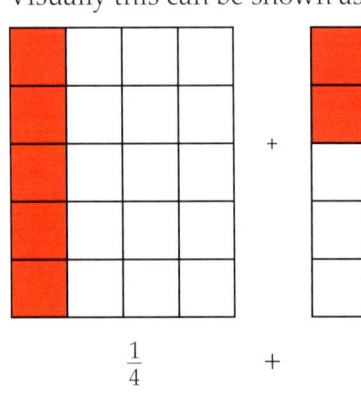 + 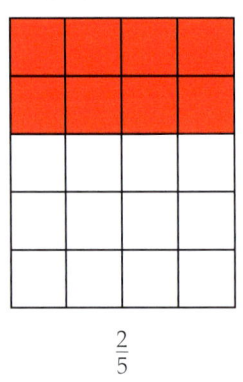 =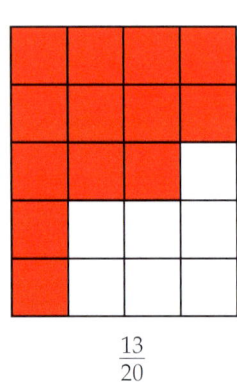

The total number of red squares is the same

$\frac{1}{4}$ + $\frac{2}{5}$ = $\frac{13}{20}$

Note that with subtraction, it is also necessary to work with fractions with a common denominator.

Exercise 15.2

1. Work out the answer to the following calculations. Show your working clearly and simplify your answers where possible.

   a $\frac{2}{5} + \frac{1}{6}$

   b $\frac{7}{12} + \frac{1}{5}$

   c $\frac{9}{14} - \frac{2}{7}$

   d $\frac{3}{13} - \frac{3}{26}$

   e $\frac{1}{8} + \frac{5}{16} - \frac{5}{24}$

   f $\frac{13}{18} - \frac{8}{9} + \frac{1}{6}$

2. Sadiq spends $\frac{1}{5}$ of his earnings on his mortgage. He saves $\frac{2}{7}$ of his earnings. What fraction of his earnings is left? Show all your working clearly.

3. A farmer uses five out of seven equal strips of his land for cereal crops and $\frac{1}{8}$ of his land for root vegetables. What fraction of his land is available for other uses? Show all your working clearly.

4. The numerators of two fractions are hidden as shown.

   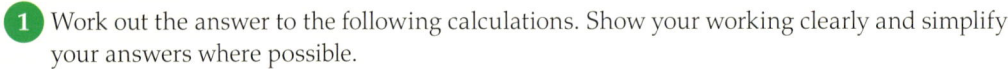

   The sum of the two fractions is $\frac{23}{40}$. Calculate the value of both numerators. Show all your working clearly.

5. The numerators of these two fractions are hidden as shown.

   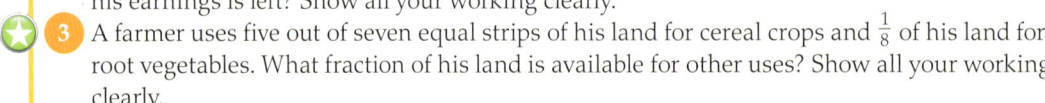

   The difference of the two fractions is $\frac{31}{60}$. Calculate the value of both numerators. Show all your working clearly.

Mixed numbers

The diagram below shows 1 unit split into quarters and $\frac{3}{4}$ of a unit.

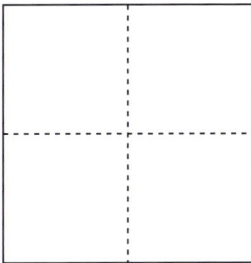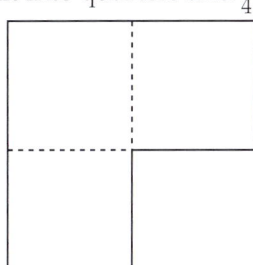

In total there are 7 quarters. As a fraction this is written as $\frac{7}{4}$. A fraction where the numerator is bigger than the denominator is known as an **improper fraction**.

However it can also be written as $1\frac{3}{4}$. A fraction that consists of a whole number and a proper fraction is called a **mixed number**.

Adding together two mixed numbers can be done in more than one way. Here one method will be explained.

> **Worked example**
>
> Add together the following pairs of mixed numbers.
>
> a $1\frac{1}{3} + 3\frac{1}{3}$
>
> Visually this can be represented as
>
>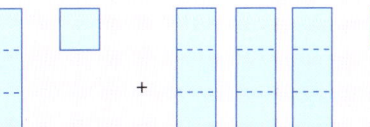
>
> By rearranging the diagram it becomes
>
>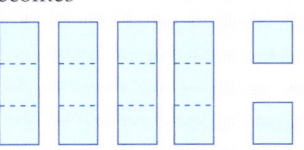
>
> > **LET'S TALK**
> > Can you think of a way of adding them together using improper fractions?
>
> The whole numbers can be added together and so can the fraction parts. Therefore, $1\frac{1}{3} + 3\frac{1}{3} = 4\frac{2}{3}$.
>
> b $2\frac{1}{4} + 5\frac{2}{3}$
>
> This can be split as before into whole numbers and fractions, i.e. $2 + 5 + \frac{1}{4} + \frac{2}{3}$.

SECTION 2

The fractional parts have different denominators, so will need to be written as equivalent fractions with the same denominator. The lowest common multiple of 4 and 3 is 12.

Therefore, $2+5+\frac{1}{4}+\frac{2}{3}$ can be written as $2+5+\frac{3}{12}+\frac{8}{12}=7\frac{11}{12}$.

c $\quad 3\frac{5}{8}+4\frac{3}{5}$

Splitting this into whole numbers and fractions gives $3+4+\frac{5}{8}+\frac{3}{5}$.

The fractional parts have different denominators and when written as equivalent fractions with the same denominator become $3+4+\frac{25}{40}+\frac{24}{40}$.

Therefore $3+4+\frac{5}{8}+\frac{3}{5}$ can be written as $3+4+\frac{25}{40}+\frac{24}{40}=7\frac{49}{40}$.

However, in this case, the fractional part is improper as the numerator is bigger than the denominator.

$\frac{49}{40}=1\frac{9}{40}$

Therefore, $7\frac{49}{40}=7+1+\frac{9}{40}=8\frac{9}{40}$.

LET'S TALK

Can you write $8\frac{9}{40}$ as an improper fraction?

Which form do you find easier to understand?

Exercise 15.3

 1 Calculate the missing number in each mixed number addition below.
   a $\quad 1\frac{\square}{7}+2\frac{1}{7}=3\frac{4}{7}$
   b $\quad 5\frac{3}{18}+3\frac{\square}{9}=8\frac{11}{18}$
   c $\quad 1\frac{8}{15}+4\frac{\square}{5}=6\frac{2}{15}$

 2 The pyramid opposite has some mixed numbers already written in some of the blocks.
   The pyramid is constructed in a way such that the fractions in the blocks in the top two rows are the sum of the two fractions in the blocks directly beneath them.
   Calculate the missing fractions.

3 The addition below sees three mixed numbers being added together.

$1\frac{1}{2}+2\frac{1}{4}+4\frac{1}{8}$

   a Describe the pattern from one fraction to the next.
   b Calculate the sum of the first five mixed numbers in this sequence.

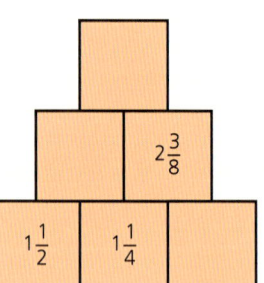

15 Fractions, decimals and percentages

 4 The pyramid below also has some mixed numbers already written in some of the blocks.

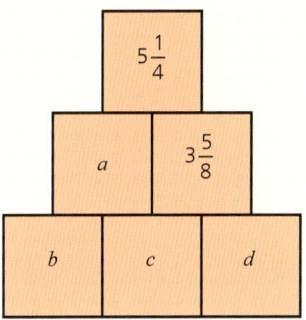

As in question 2 each block has a fraction which is the sum of the two fractions directly beneath it.
a Which block(s) of *a*, *b*, *c* or *d* can only have one possible fraction as an answer? Justify your answer.
b Give a possible fraction for the block(s) where more than one answer is possible.

Multiplication and division of fractions

Multiplication

In everyday language, to say 'half of' an amount is the same as multiplying that amount by a half or dividing by 2.

For example: Half of $20 can be written as either $20 \times \frac{1}{2}$ or $20 \div 2$.

Similarly, to say 'a quarter' of an amount is the same as multiplying that amount by a quarter or dividing by 4.

For example: A quarter of $\frac{1}{2}$ can be written as either $\frac{1}{2} \times \frac{1}{4}$ or $\frac{1}{2} \div 4$.

The answer to the calculation $\frac{1}{2} \div 4$ can be visualised below.

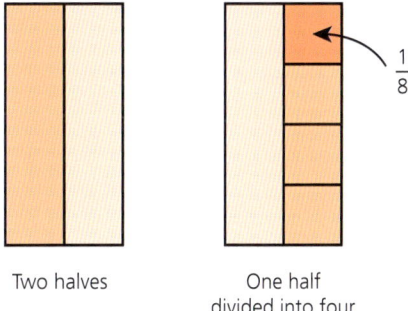

Two halves   One half divided into four

In the diagrams above it can be seen that $\frac{1}{2} \div 4 = \frac{1}{8}$. Therefore $\frac{1}{2} \times \frac{1}{4} = \frac{1}{8}$ too.

SECTION 2

To multiply fractions together simply multiply the numerators together and multiply the denominators together.

For example: $\frac{1}{2} \times \frac{1}{4} = \frac{1 \times 1}{2 \times 4} = \frac{1}{8}$.

Worked examples

1. Multiply the following fractions $\frac{3}{7} \times \frac{4}{5}$.

   Multiplying the numerators together and multiplying the denominators together gives:

   $$\frac{3}{7} \times \frac{4}{5} = \frac{3 \times 4}{7 \times 5} = \frac{12}{35}$$

   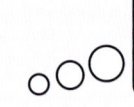

   Remember to check whether the final answer can be simplified.

2. Multiply the following fractions $\frac{3}{8} \times \frac{4}{7}$.

   Multiplying the numerators together and multiplying the denominators together gives:

   $$\frac{3}{8} \times \frac{4}{7} = \frac{3 \times 4}{8 \times 7} = \frac{12}{56}$$

   The answer can be simplified as both numerator and denominator are divisible by 4. Therefore $\frac{12}{56} = \frac{3}{14}$.

   It is possible to make the calculation easier by simplifying the fractions before multiplication.

   It can be seen that 4 is a common factor of one of the numbers in the numerator and also one of the numbers in the denominator.

   $$\frac{3}{\cancel{8}_2} \times \frac{\cancel{4}^1}{7} = \frac{3}{2} \times \frac{1}{7} = \frac{3}{14}$$

3. Multiply the following fractions $\frac{8}{15} \times \frac{5}{32}$.

   The multiplication is difficult to do without simplifying first.

   Here both numerator and denominator are divisible by 5.

   $$\frac{8}{\cancel{15}_3} \times \frac{\cancel{5}^1}{32}$$

   But both numerator and denominator are also divisible by 8.

   $$\frac{\cancel{8}^1}{15} \times \frac{5}{\cancel{32}_4}$$

   Therefore $\frac{8}{15} \times \frac{5}{32}$ can be simplified first to $\frac{1}{3} \times \frac{1}{4} = \frac{1}{12}$.

LET'S TALK

$\frac{8}{15} \times \frac{5}{32}$ has been simplified to $\frac{1}{3} \times \frac{1}{4}$ but $\frac{8}{15} \neq \frac{1}{3}$ and $\frac{5}{32} \neq \frac{1}{4}$ so why is $\frac{8}{15} \times \frac{5}{32}$ equivalent to $\frac{1}{3} \times \frac{1}{4}$?

15 Fractions, decimals and percentages

Division

Earlier it was shown that $\frac{1}{2} \div 4 = \frac{1}{2} \times \frac{1}{4}$ therefore a division of fractions can be written as a multiplication of fractions and vice versa.

Note that the **reciprocal** of 4 is $\frac{1}{4}$.

Therefore dividing by a number or fraction is the same as multiplying by its **reciprocal**.

> **KEY INFORMATION**
>
> The reciprocal of a number is 1 divided by that number.
>
> Therefore, the reciprocal of 3 is $\frac{1}{3}$.
>
> The reciprocal of $\frac{2}{3}$ would be $\frac{1}{\frac{2}{3}}$ which is more commonly written as $\frac{3}{2}$.

> **LET'S TALK**
>
> Why is $\frac{1}{\frac{2}{3}}$ the same as $\frac{3}{2}$?
>
> Can you show this visually?

Worked examples

1. Work out the following division, $\frac{1}{3} \div \frac{2}{5}$.

   The division can be written as a multiplication by multiplying the first fraction by the reciprocal of the second one. Therefore:

   $$\frac{1}{3} \div \frac{2}{5} = \frac{1}{3} \times \frac{5}{2} = \frac{5}{6}$$

2. Work out the following division, $\frac{5}{8} \div \frac{3}{4}$.

   Changing the division to a multiplication becomes $\frac{5}{8} \times \frac{4}{3}$.

   As before this can be simplified before multiplication as 4 is a factor of both the numerator and denominator.

   $$\frac{5}{\underset{2}{\cancel{8}}} \times \frac{\overset{1}{\cancel{4}}}{3} = \frac{5}{2} \times \frac{1}{3} = \frac{5}{6}$$

   > As with example **1** above, by dividing by $\frac{3}{4}$ which is less than 1, the answer will be bigger than $\frac{5}{8}$, so $\frac{5}{6}$ is bigger than $\frac{5}{8}$.

> **KEY INFORMATION**
>
> Note that when dividing by a fraction less than 1, the answer is bigger than the original fraction.
>
> That is $\frac{5}{6}$ is bigger than $\frac{1}{3}$.

3. Work out the following division, $\frac{8}{11} \div \frac{2}{5}$.

   Changing the division to a multiplication becomes $\frac{8}{11} \times \frac{5}{2}$.

   This can be simplified to $\frac{\overset{4}{\cancel{8}}}{11} \times \frac{5}{\underset{1}{\cancel{2}}}$.

   Therefore $\frac{8}{11} \times \frac{5}{2} = \frac{4}{11} \times \frac{5}{1} = \frac{20}{11}$.

   As the answer is an improper fraction it can be written as a mixed number as $1\frac{9}{11}$.

SECTION 2

Exercise 15.4

1. Multiply the following fractions, simplifying your answer where possible.
   a $\frac{6}{7} \times \frac{2}{5}$
   b $\frac{3}{10} \times \frac{1}{6}$
   c $\frac{2}{21} \times \frac{7}{40}$
   d $\frac{3}{8} \times \frac{4}{15}$
   e $\frac{9}{16} \times \frac{8}{27}$

2. Eight cards are shown below. Match each fraction with its reciprocal.

   $\frac{1}{3}$ $\frac{3}{2}$ $\frac{3}{5}$ 3 $\frac{1}{5}$ $\frac{5}{3}$ 5 $\frac{2}{3}$

3. Divide the following fractions, simplifying your answer where possible.
   a $\frac{1}{6} \div 2$
   b $\frac{2}{9} \div 3$
   c $\frac{7}{10} \div \frac{7}{8}$
   d $\frac{4}{15} \div \frac{8}{25}$
   e $\frac{5}{12} \div \frac{3}{8}$

4. Without working out the following divisions, decide whether each statement is true or false. Justify your answer each time.
   a $\frac{2}{3} \times 3 > \frac{2}{3}$
   b $\frac{3}{7} \times \frac{1}{3} < \frac{3}{7}$
   c $\frac{2}{5} \div 4 < \frac{2}{5}$
   d $\frac{5}{9} \div \frac{1}{2} > \frac{5}{9}$
   e $\frac{7}{10} \div \frac{7}{10} \neq \frac{7}{10}$

   KEY INFORMATION
   = means equal to
   ≠ means not equal to
   > means more than
   < means less than

5. Look at the four calculations below.
   $\frac{3}{8} \times \frac{2}{7}$ $\frac{1}{4} \div \frac{7}{3}$ $\frac{2}{3} \times \frac{7}{8}$ $\frac{3}{7} \times \frac{2}{8}$
   a Which of the four calculations will produce a different answer? Justify your answer.
   b Write another fraction calculation which will produce the same answer as the other three. Justify your choice.

Manipulating fractions and decimals

With the multiplication of fractions covered earlier, we saw that $\frac{8}{15} \times \frac{5}{32}$ could be simplified before multiplication by looking for common factors

For example, $\frac{\cancel{8}^1}{\cancel{15}_3} \times \frac{\cancel{5}^1}{\cancel{32}_4} = \frac{1 \times 1}{3 \times 4} = \frac{1}{12}$

15 Fractions, decimals and percentages

Searching for ways to simplify calculations makes the calculation easier.

Worked examples

1. Work out the following without a calculator.

   $0.5 \times 9 \times 10$

   It is possible to carry out the multiplication in the order it is written.

   So, as $0.5 \times 9 = 4.5$ the calculation becomes $4.5 \times 10 = 45$.

   However, as multiplication is **commutative**, it can be done in a different order.

   You may find it easier to do the multiplication like this: $0.5 \times 10 \times 9$.

   As $0.5 \times 10 = 5$ the calculation becomes $5 \times 9 = 45$.

2. Work out the following without a calculator.

   $18 \times 9 \times \frac{1}{3}$

   As multiplication is commutative, the order of the multiplication does not matter. 18×9 is not easy to do in your head. Here changing the order to $\frac{1}{3} \times 18 \times 9$ makes the calculation significantly easier.

   $\frac{1}{3} \times 18 = 6$, therefore the calculation becomes $6 \times 9 = 54$.

3. Work out the following without a calculator.

   5.2×12

   This can be written as $5.2 \times (10 + 2)$.

   The 5.2 is now multiplying both numbers inside the bracket.

   So, 5.2×10 and 5.2×2.

   The calculation therefore becomes $52 + 10.4 = 62.4$.

4. Arrange the following fractions in order of size from smallest to largest.

   $\frac{9}{16} \quad \frac{5}{8} \quad \frac{3}{7}$

   There are many ways of tackling this question. The Let's talk box opposite considers another method.

   Just by looking at the fractions it can be seen that $\frac{3}{7}$ must be the smallest as it is the only one less than half.

   To compare $\frac{9}{16}$ and $\frac{5}{8}$, write them as equivalent fractions with the same denominator. 16 is the lowest common denominator, therefore $\frac{5}{8}$ can be written as the equivalent fraction $\frac{10}{16}$.

   In order the fractions are therefore $\frac{3}{7} \quad \frac{9}{16} \quad \frac{10}{16}$ or to give each fraction in their original form, the order is $\frac{3}{7} \quad \frac{9}{16} \quad \frac{5}{8}$.

KEY INFORMATION

Multiplication is commutative, which means that changing the order of the calculation will not affect the answer, e.g. $3 \times 4 = 4 \times 3$.

KEY INFORMATION

You saw in Unit 14 that an algebraic expression like $3(a+4)$ can be expanded to $3a+12$ as the 3 outside the bracket multiplies both terms inside the bracket. The same can be applied to numbers.

LET'S TALK

What is the lowest common multiple of 16, 8 and 7?

Why is finding this out not an efficient method for answering this question?

SECTION 2

Exercise 15.5

1. a The rectangle below has the dimensions shown.

   12 cm

   7.1 cm

   Explain why the area of the rectangle can be calculated by the calculation $(7.1 \times 10) + (7.1 \times 2)$.

   b Work out the area of the rectangle below. Show your working clearly. Is there a different calculation you could have used? Which calculation is easiest to use?

   15.3 cm

   8 cm

2. Two athletes drink different fractions of liquid from identical water bottles. Athlete A drinks $\frac{11}{24}$ of his bottle, whilst athlete B drinks $\frac{9}{16}$ of her bottle. Show all your working clearly.
   Which athlete has drunk the most? Justify your answer.

3. Work out the following.
   a $15 \times 12 \times \frac{1}{4}$
   b 18.3×9
   c $0.25 \times 17 \times 24$
   d $18 \times 13 \times \frac{1}{6}$
   e 15.1×22
   f $\frac{3}{4} \times \frac{5}{8} + \frac{3}{16}$

4. An artist wants to mix two paints X and Y together to create a new colour. He knows that to get the colour he wants he must add between $\frac{7}{18}$ of tube X to $\frac{5}{9}$ of tube Y.
   Give *two* different fractions of tube X that he can add.

5 Three identical fruit juice dispensers P, Q and R in a hotel restaurant have the same amount of juice at the start of breakfast.

The fraction of juice left at the end of breakfast in each dispenser is given below.

P	$\frac{7}{12}$
Q	$\frac{17}{24}$
R	$\frac{19}{36}$

a Which dispenser has the most juice left at the end? Justify your answer.
b Which dispenser has the least juice left at the end? Justify your answer.

6 Order the following fractions from the smallest to the largest

$3\frac{1}{3}$ $\frac{9}{4}$ $\frac{11}{5}$ $\frac{15}{8}$ $2\frac{2}{5}$

7 A family is driving to a small village. The distance of the village from their current position is between $18\frac{1}{5}$ km and $18\frac{2}{7}$ km away.
Can the village be $18\frac{12}{35}$ km away? Justify your answer.

 Now you have completed Unit 15, you may like to try the Unit 15 online knowledge test if you are using the Boost eBook.

16 Probability and outcomes

- Identify all the possible mutually exclusive outcomes of a single event, and recognise when they are equally likely to happen.
- Understand how to find the theoretical probabilities of equally likely outcomes.

Mutually exclusive outcomes

When flipping a coin, only two events are possible, getting a head or getting a tail. They cannot happen at the same time. Outcomes which cannot happen at the same time are called **mutually exclusive**.

LET'S TALK
Are each of the outcomes equally likely? How likely are they?

Worked examples

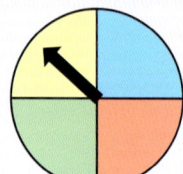

1 A spinner has four colours as shown.
   a List all the possible outcomes.
      Yellow, Blue, Red and Green
   b Explain whether the outcomes are mutually exclusive.
      Yes, they are mutually exclusive outcomes, because they cannot occur at the same time. For example, the spinner cannot land on both yellow and red at the same time.

2 a List five different weather outcomes possible for tomorrow.
      Rain, Wind, Sun, Warm, Cold
   b Explain whether the weather outcomes are mutually exclusive
      No, they are not mutually exclusive because they can happen at the same time, for example a day can be both rainy and cold, or windy, sunny and warm at the same time.

3 A school hockey team plays a match. The probability of their winning is estimated to be 0.5 and the probability of losing 0.3. Explain why the probability of the team drawing must be 0.2.
   In a match the only outcomes are winning, losing or drawing and these outcomes are mutually exclusive (i.e. a team cannot both win and lose a match).
   The total probability of all possible outcomes is 1, therefore:
   $1 - 0.5 - 0.3 = 0.2$

LET'S TALK
Are each of the outcomes here equally likely?

How could you calculate each of their probabilities?

16 Probability and outcomes

Exercise 16.1

1. A board game states that to start playing, a player must roll a 6 with the dice.
   a. What are the possible outcomes for a player when rolling a dice?
   b. What is the probability that the player rolling the dice will start the game on their first attempt?

2. The diagrams below show three different spinners.

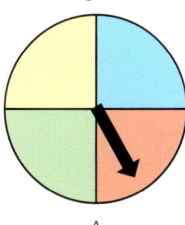 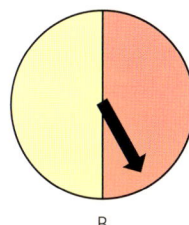 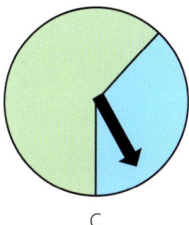

   A   B   C

   a. Which spinner(s) show equally likely outcomes? Justify your answer.
   b. For the spinner(s) with equally likely outcomes, state the probability of getting each colour.

3. An archer fires an arrow at the target to the right.
   a. The archer hits the target with his first arrow. List the possible outcomes.
   b. Are the outcomes you listed in part (a) equally likely? Justify your answer.
   c. Are the outcomes listed in part (a) mutually exclusive? Justify your answer.
   d. The archer fires several arrows at the target. They all hit the target. He works out that his probability of hitting the blue ring was 0.2 and the probability of hitting the red ring was 0.4. How likely was he to hit the black ring compared to the red ring? Justify your answer.

4. A school offers its students the option of two sports to play, volleyball (V) and basketball (B). Below is a Venn diagram showing the probability of students who choose each.

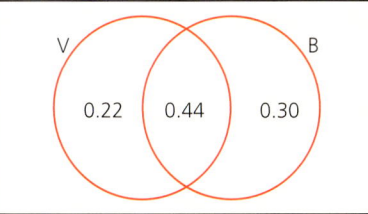

   a. What is the probability of a student not playing either sport?
   b. A student is chosen at random. What is the probability that they play volleyball?
   c. Are the two options mutually exclusive? Justify your answer.

SECTION 2

 5 In a school sports day, the number of students taking part in three of the field events are shown in the two-way table below.

	Long jump	High jump	Javelin
Girls	6	2	5
Boys	8	0	7

a A student is picked at random. What is the probability of picking one who does the long jump?
b List the different possible outcomes displayed in the table.
c Are the different outcomes listed in (b) above equally likely? Justify your answer.
d Are any of the outcomes mutually exclusive? Justify your answer.

LET'S TALK
Why is the word **theoretical** used to describe the probability?

You will have already seen in Unit 10 that calculating the **theoretical probability** of an event happening depends on the number of equally likely outcomes. For example, the theoretical probability of getting a head when flipping a coin is $\frac{1}{2}$ as only one outcome is a **head** out of two possible outcomes, **heads** or **tails**.

Calculating the theoretical probability of equally likely outcomes can therefore be written as a formula:

$$\text{Theoretical probability} = \frac{\text{Number of successful outcomes}}{\text{Total number of equally likely outcomes}}$$

As we were looking at **heads**, getting a **head** is considered a **successful outcome**.

Worked examples

1 Three cards, with the numbers 1, 2 and 3 are turned over and shuffled as shown.

A card is then picked at random.
a Calculate the theoretical probability of picking the 2.

$$\text{Theoretical probability} = \frac{\text{Number of successful outcomes}}{\text{Total number of equally likely outcomes}}$$

$$= \frac{1}{3}$$

16 Probability and outcomes

LET'S TALK
Why is $\frac{1}{3}$ subtracted from 1?

b Calculate the theoretical probability of not picking the 2.
There are two ways in which this can be calculated.
As there are two other numbers, other than 2, and they are as equally likely to be picked, the probability of not picking 2 is $\frac{2}{3}$.
Alternatively as the probability of picking 2 is $\frac{1}{3}$, then the probability of not picking 2 is calculated as $1 - \frac{1}{3} = \frac{2}{3}$.

2 A biased dice is rolled.
   a Explain why the theoretical probability of getting a 6 is not likely to be $\frac{1}{6}$.
   The numbers on a biased dice are not equally likely, therefore the probability of getting a 6 cannot be calculated using the formula for theoretical probability.

LET'S TALK
Why is the dice rolled so many times for this calculation? Wouldn't rolling the dice only 10 times be enough?

   b How can the probability of getting a 6 on a biased dice be calculated?
   This will have to be calculated by carrying out an experiment. The dice can be rolled 1000 times for example and the number of 6's recorded. This will give an idea of how likely it is to roll a 6 in future.

Exercise 16.2

 1 Carla is playing a game of football with her team. She says that as there are only three outcomes – win, lose or draw – the probability of her team winning is $\frac{1}{3}$.
Explain whether Carla is right or wrong.

 2 Three different medicine manufacturers test the effectivess of their own brand of medicine for curing patients of a certain illness.
The results of the tests are as follows:
- Manufacturer A tested 1 patient and the patient recovered. They claim that their medicine is effective in 100% of cases.
- Manufacturer B tested 100 patients and 60 recovered. They claim that their medicine is effective in 60% of cases.
- Manufacturer C tested 10 000 patients and 5000 recovered. They claim that their medicine is effective in 50% of cases.

Which manufacturer's claim is the most reliable? Justify your answer.

SECTION 2

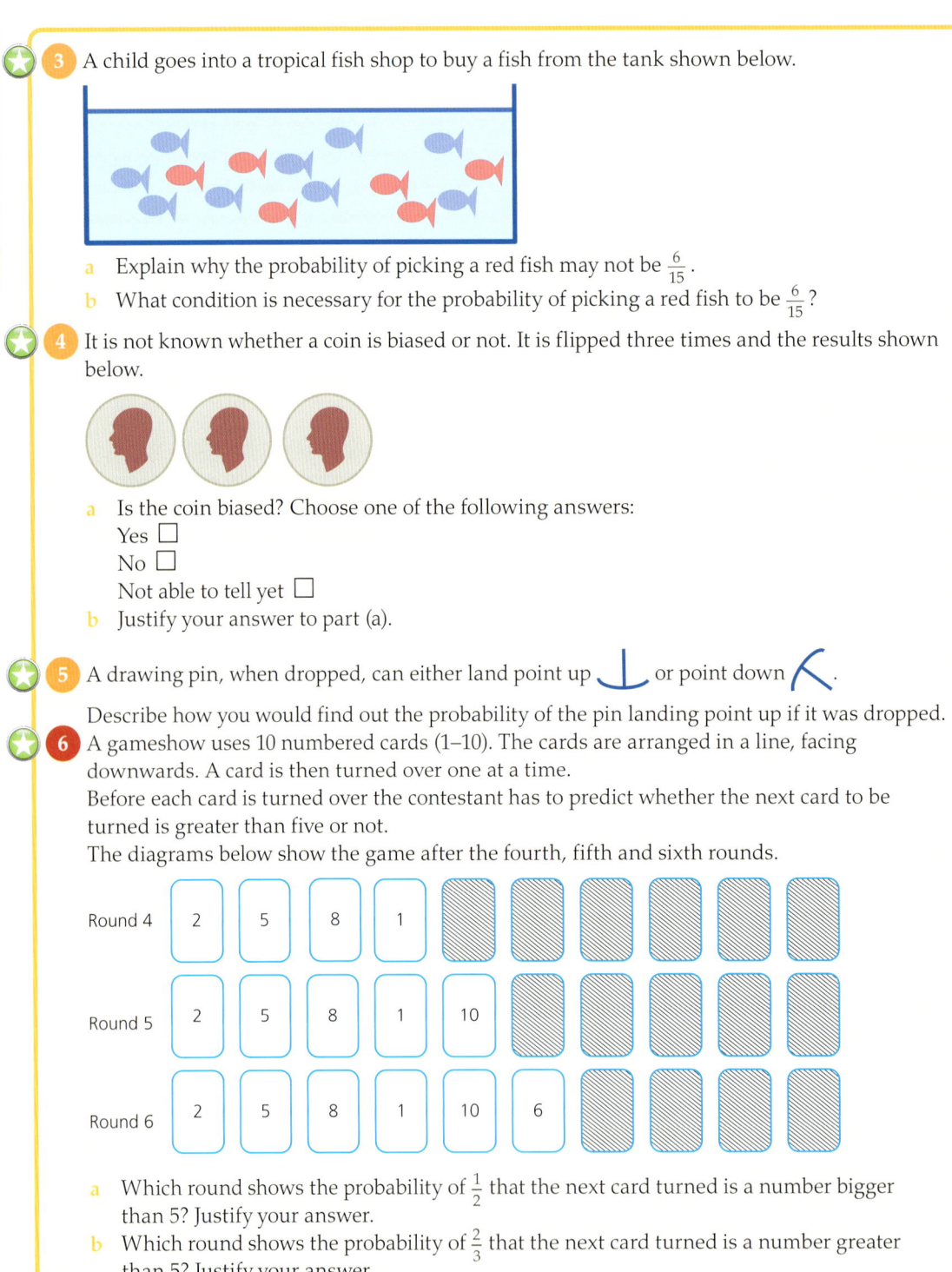

3 A child goes into a tropical fish shop to buy a fish from the tank shown below.

   a Explain why the probability of picking a red fish may not be $\frac{6}{15}$.
   b What condition is necessary for the probability of picking a red fish to be $\frac{6}{15}$?

4 It is not known whether a coin is biased or not. It is flipped three times and the results shown below.

   a Is the coin biased? Choose one of the following answers:
      Yes ☐
      No ☐
      Not able to tell yet ☐
   b Justify your answer to part (a).

5 A drawing pin, when dropped, can either land point up or point down.
   Describe how you would find out the probability of the pin landing point up if it was dropped.

6 A gameshow uses 10 numbered cards (1–10). The cards are arranged in a line, facing downwards. A card is then turned over one at a time.
   Before each card is turned over the contestant has to predict whether the next card to be turned is greater than five or not.
   The diagrams below show the game after the fourth, fifth and sixth rounds.

   Round 4: 2, 5, 8, 1
   Round 5: 2, 5, 8, 1, 10
   Round 6: 2, 5, 8, 1, 10, 6

   a Which round shows the probability of $\frac{1}{2}$ that the next card turned is a number bigger than 5? Justify your answer.
   b Which round shows the probability of $\frac{2}{3}$ that the next card turned is a number greater than 5? Justify your answer.

> Now you have completed Unit 16, you may like to try the Unit 16 online knowledge test if you are using the Boost eBook.

17 Angle properties

- Draw parallel and perpendicular lines and quadrilaterals.
- Derive the property that the sum of the angles in a quadrilateral is 360°.
- Know that the sum of the angles around a point is 360°.
- Recognise the properties of angles on parallel, perpendicular and intersecting lines.

Angle construction

An angle is a measure of turn. It can be drawn using a protractor or an angle measurer. The units of turn are degrees (°).

Worked examples

1. Measure the angle drawn below.

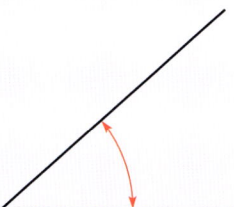

- Place the protractor over the angle as shown.

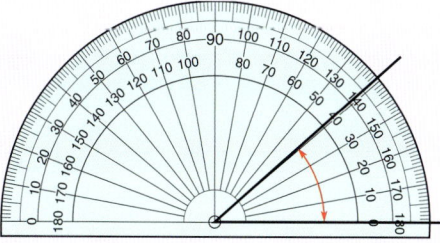

- Align the 0° with one of the lines drawn
- Look at the scales to see which one you will use. In this case it is the inner scale which starts at 0°. The angle is 41°.

2. Draw an angle of 120°.
   - First draw a straight line about 6 cm long.
   - Place the protractor on the line so that the central cross hair is on one of the end points of the line. Make sure the line lines up with the 0° on the protractor.

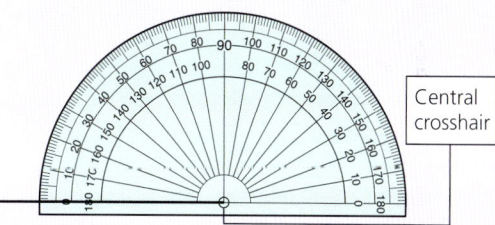

Central crosshair

- Decide which scale to use. (In this case it is the outer scale as it starts at 0°.)
- Mark where the protractor reads 120°.
- Join the mark to the end of the line.

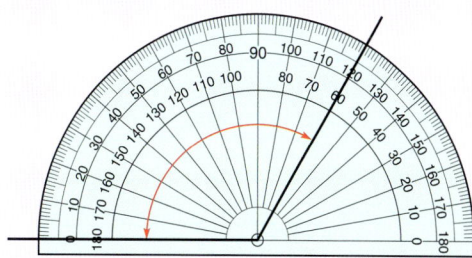

135

SECTION 2

Exercise 17.1

1 For each of the angles shown:
   i) estimate its size
   ii) measure it and check how good your estimate was.
      Aim for your estimate to be within 10° of the actual angle.

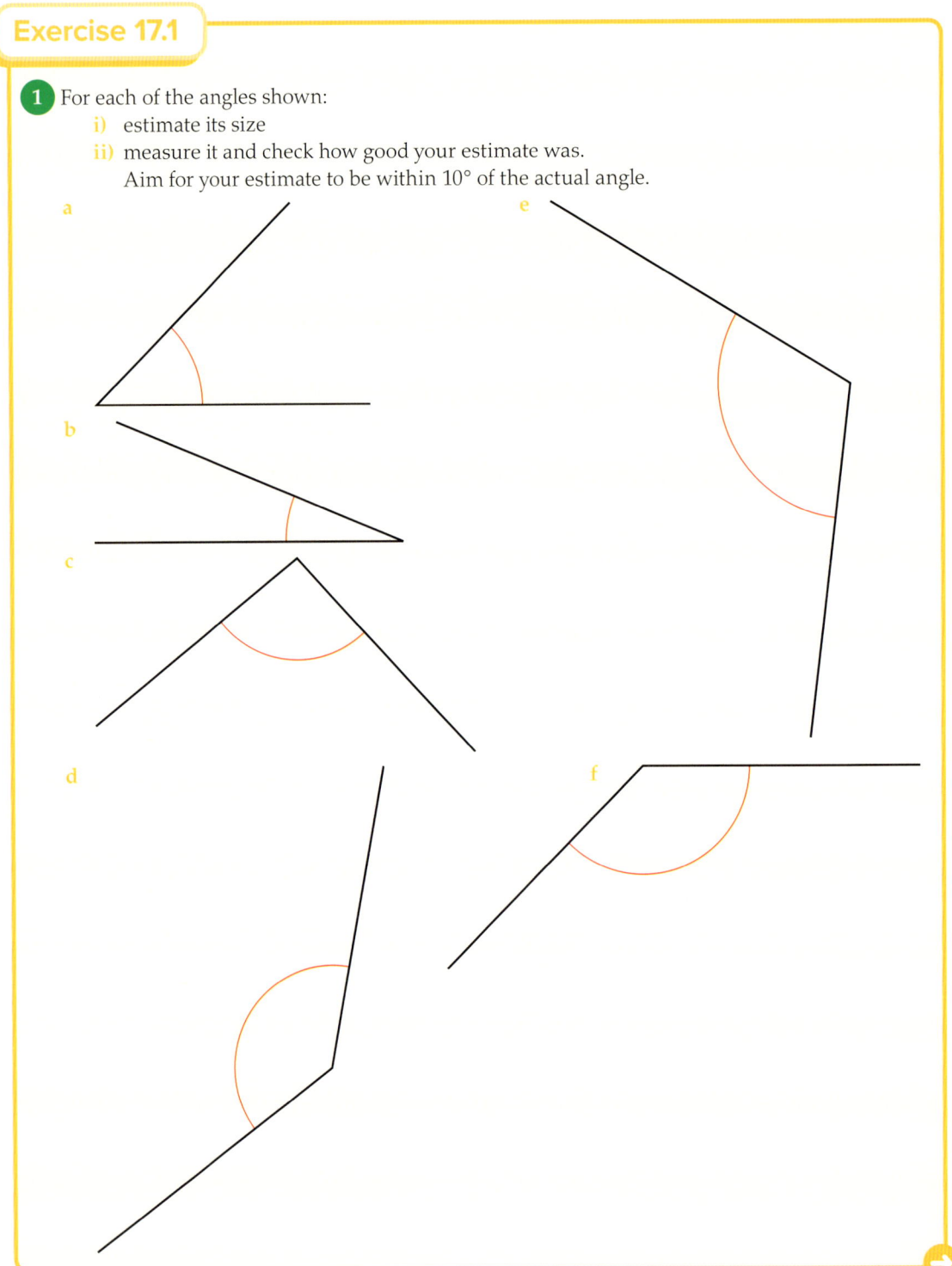

17 Angle properties

2 Measure the angles in the diagrams below.

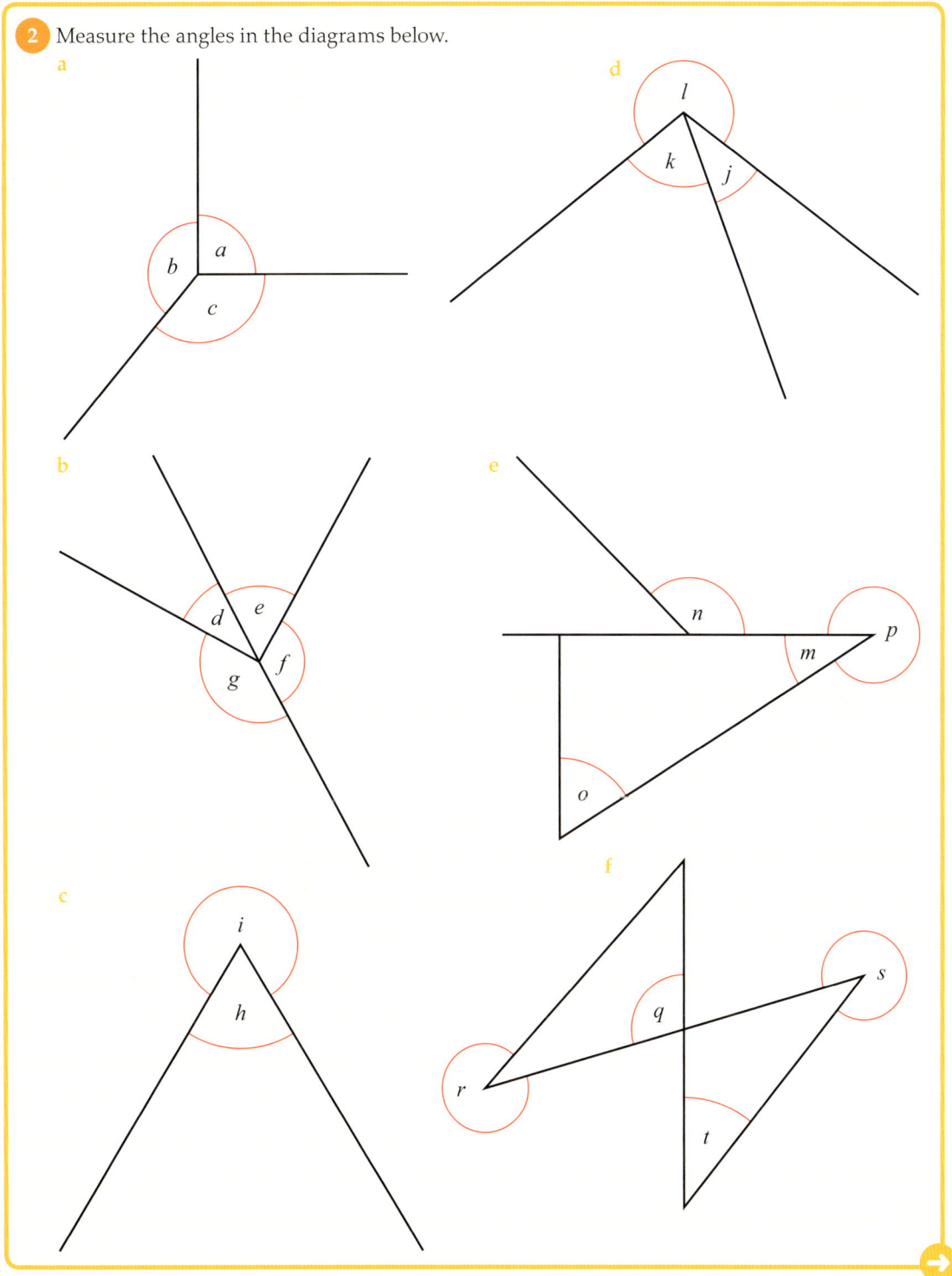

SECTION 2

3) In the following question:
   i) draw by estimating (without a protractor) angles of the following size
   ii) alongside your first diagram, draw the same angles with a protractor.
   Compare your answers. How good were your estimates?

   a 20°   e 157°   i 311°
   b 45°   f 172°   j 283°
   c 90°   g 14°   k 198°
   d 120°   h 205°   l 352°

Constructing simple geometric figures

Quadrilaterals

Constructing a quadrilateral accurately requires careful use of a ruler and a protractor.

> **Worked example**
>
> Construct a parallelogram ABCD of side lengths 8 cm and 4 cm with internal angles of 108° and 72°.
> - Draw a line AB, 8 cm long.
>
>   B ———————————— A
>   8 cm
>
> - Place the protractor on B and mark off an internal angle of 108°.

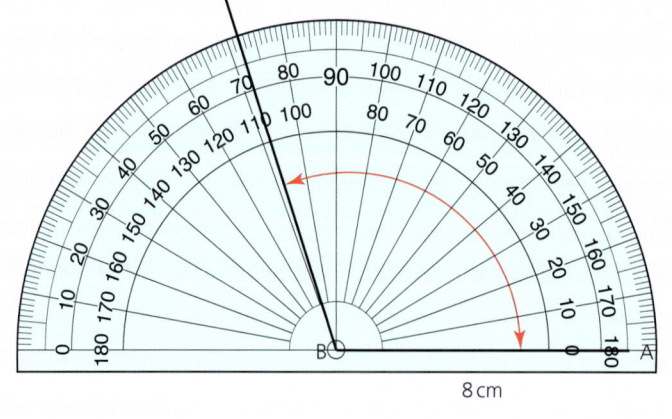

Remember a parallelogram has two pairs of parallel sides and adjacent angles add up to 180°.

Remember to leave enough space above the line to draw the rest of the parallelogram.

138

17 Angle properties

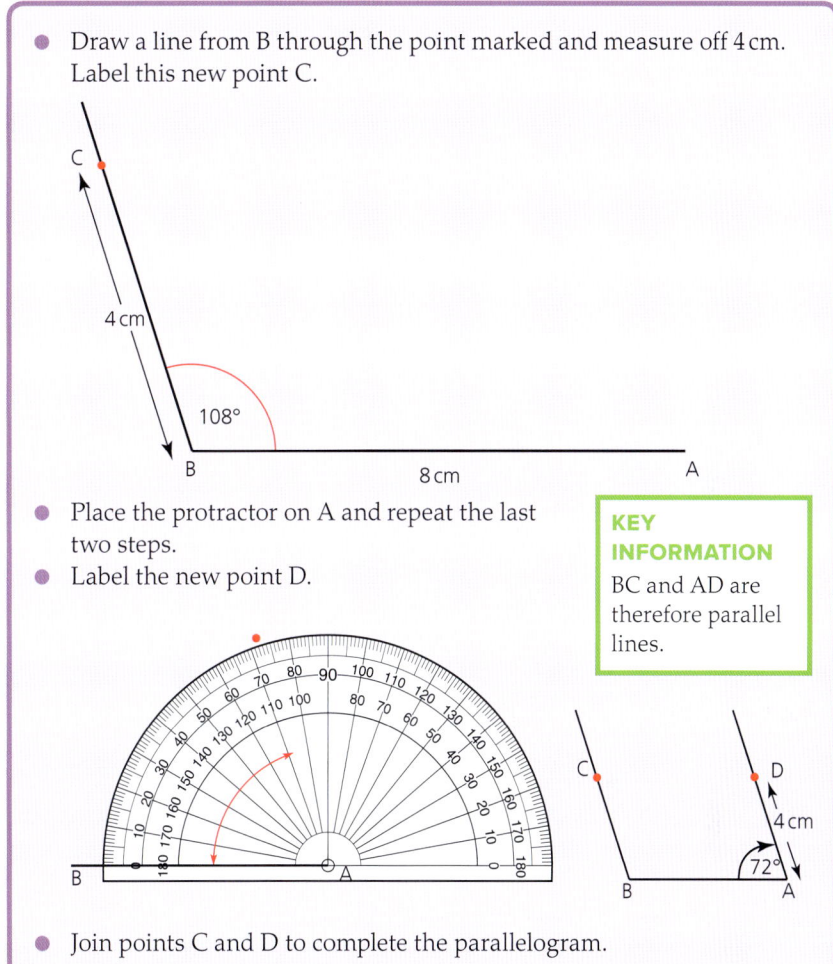

- Draw a line from B through the point marked and measure off 4 cm. Label this new point C.
- Place the protractor on A and repeat the last two steps.
- Label the new point D.

KEY INFORMATION
BC and AD are therefore parallel lines.

KEY INFORMATION
As the angles add up to 180° the different scales on the protractor can be used, but the position of the point remains the same. Here the angle of 72° is shown.

- Join points C and D to complete the parallelogram.

Squares and rectangles

Squares and rectangles too can be constructed. To do this both a ruler and a set square are needed. The set square is used as it provides a right angle with which to draw lines **perpendicular** (at right angles) to each other.

There are two different types of set square. Both have a right angle, but one can draw angles of 30° and 60°, whilst the other can be used to draw angles of 45°.

139

SECTION 2

> **Worked example**
>
> Construct a square of side length 6 cm.
> - Draw a line 6 cm long.
> - Place a set square at the end of the line, making sure that one of the perpendicular sides rests on the line drawn like this:
>
>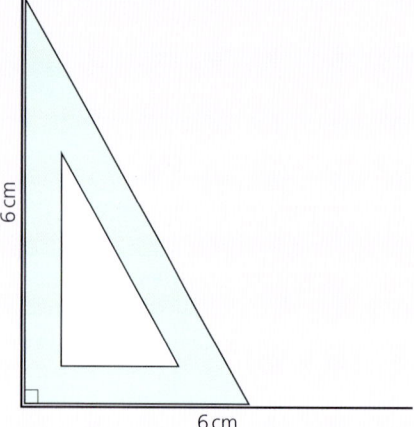
>
> - Repeat this for the remaining two sides.

Exercise 17.2

1 Using a ruler and set squares only, construct the following shapes:

a

b

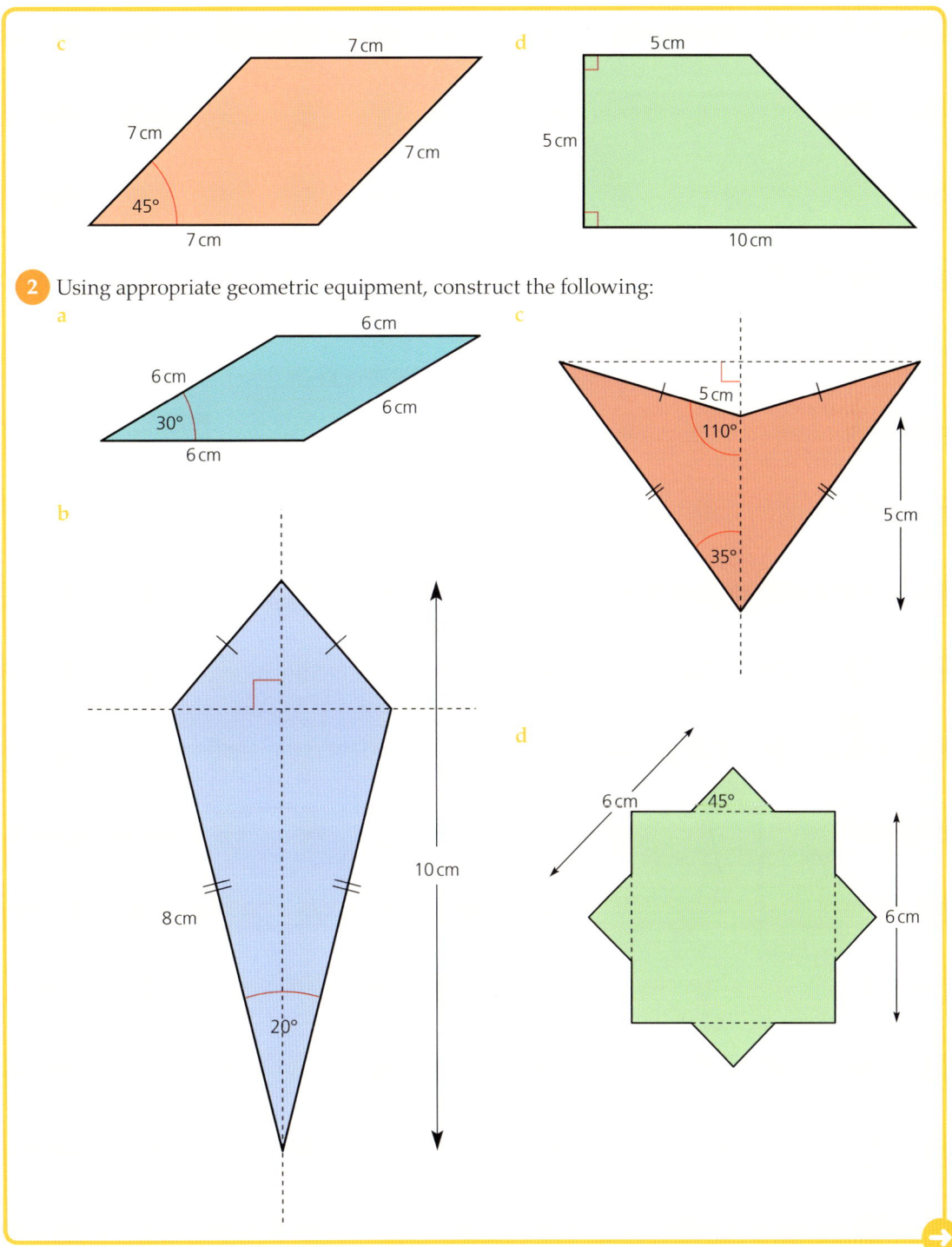

2. Using appropriate geometric equipment, construct the following:

SECTION 2

 3 Two types of set square are shown below:

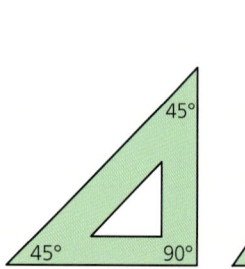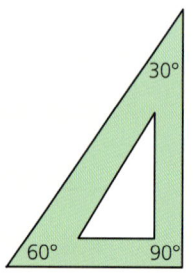

a An architect says that she can draw an angle of 75° using these two set squares. Show how this can be done.
b She also says she can draw an angle of 15°. Show how this can be done.
c State two **obtuse** angles she could also draw using her set squares. Justify your answer.

KEY INFORMATION
An obtuse angle is between 90° and 180°.

Angle properties of quadrilaterals

You will already know that the three angles of any triangle always add up to 180°.

A similar rule also exists for the sum of the four angles of any quadrilateral.

Any quadrilateral can be split in to two triangles as shown:

LET'S TALK
Using a similar method can you work out what the five interior angles of any pentagon must add up to?

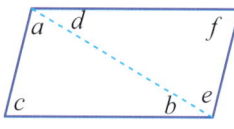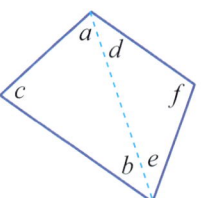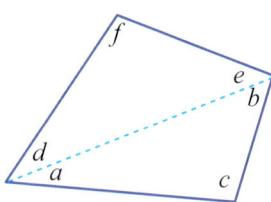

In each case $a+b+c=180°$ and $d+e+f=180°$.

As the angles a, b, c, d, e and f make up the four angles of each quadrilateral, the conclusion must be that the angles of any quadrilateral add up to 360°.

The four angles of any quadrilateral always add up to 360°

17 Angle properties

Exercise 17.3

In each of the questions below calculate the size of the angles labelled with letters, giving a reason for your answer each time.

1. Calculate the size of angle a. Justify your answer.

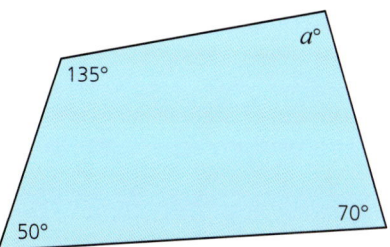

2. a What is the name of the following quadrilateral?

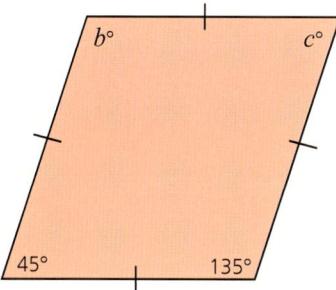

   b Calculate the size of angles b and c. Justify your answers.

3. a Name the quadrilateral below. Justify your answer.

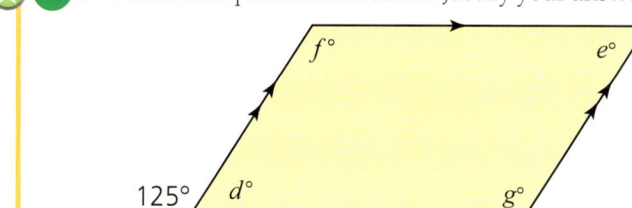

   b Calculate each of the angles d, e, f and g. Justify your answers.

SECTION 2

4 A kite is shown.

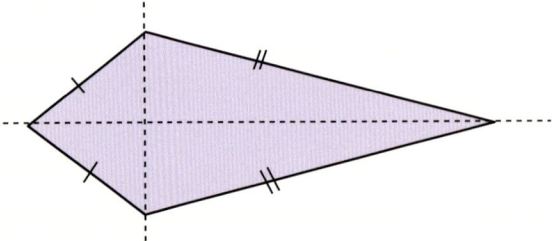

a Which of the following properties are true for a kite?
- It has one pair of equal opposite angles.
- It has two pairs of equal opposite angles.
- Its diagonals intersect at right angles.
- It has two pairs of adjacent sides of the same length.
- It has one line of reflective symmetry.
- It has two lines of reflective symmetry.
- It has no rotational symmetry.
- It has a rotational symmetry of order 2.
- Its internal angles add up to 360°.

b Calculate each of the angles a, b and c.
Justify your answers using some of the properties you identified in part (a) above.

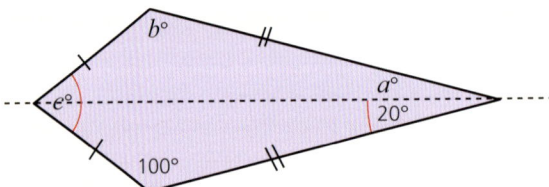

5 a Describe five properties of a parallelogram.
b Calculate the size of the angles d, e and f in the parallelogram below. Justify your answers using some of the properties you identified in part (a).

6 The diagram below shows an isosceles trapezium ABCD inside a parallelogram ABED.

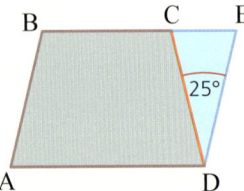

KEY INFORMATION
An isosceles trapezium has one line of reflective symmetry.

Angle CDE is 25°.
Calculate the size of each of the angles of the trapezium. Justify your answers fully.

144

17 Angle properties

Angles around a point

You will already know that the angles on a straight line add up to 180°.

Therefore $a+b=180°$ and similarly $c+d+e=180°$.

If the two diagrams are joined around a point, it becomes:

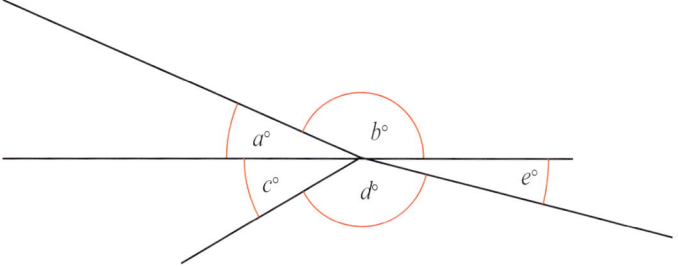

Therefore as $180°+180°=360°$, $a+b+c+d+e=360°$.

Angles around a point add up to 360°

> **Worked example**
>
> Calculate the size of the angle x in the diagram below.
>
>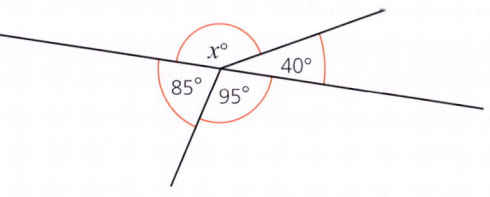
>
> $x+40+95+85=360$
>
> $x+220=360$
>
> $x=360-220=140°$

145

Exercise 17.4

1. Find the size of the unknown angle in each of these questions:

a) 120°, 130°, a

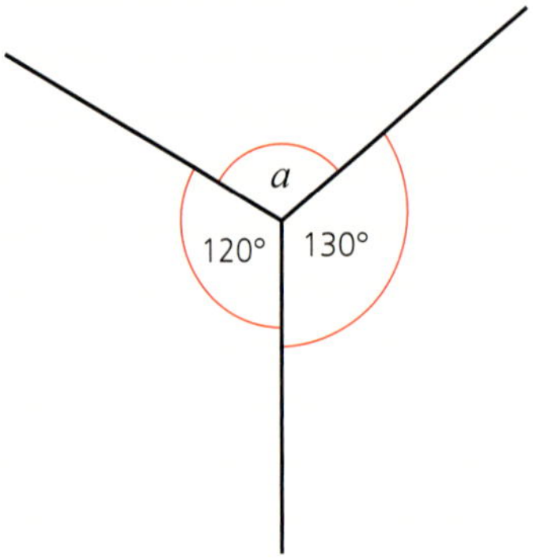

b) 125°, right angle, b

c) 85°, 220°, c

d) 125°, 55°, 85°, d

e) 95°, 120°, 45°, e

f) 58°, 62°, 115°, f

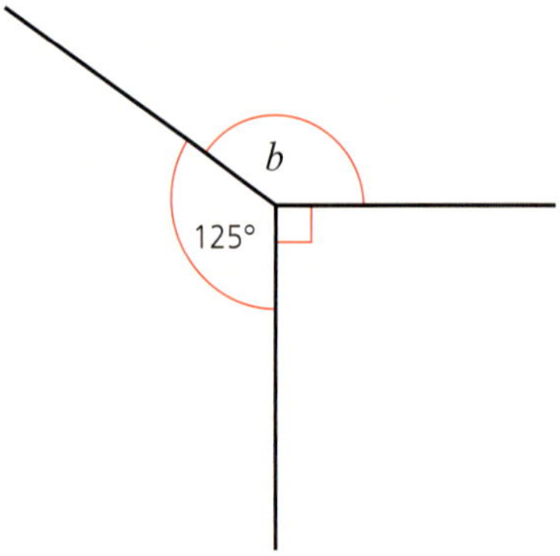

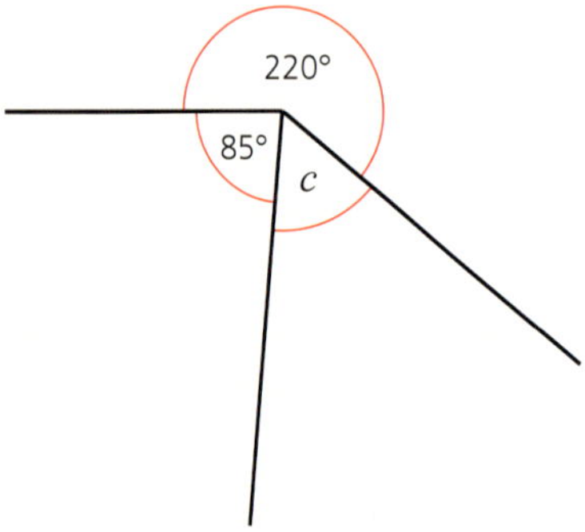

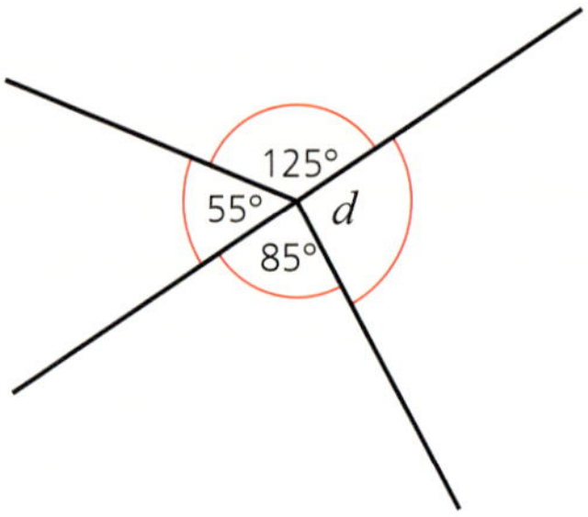

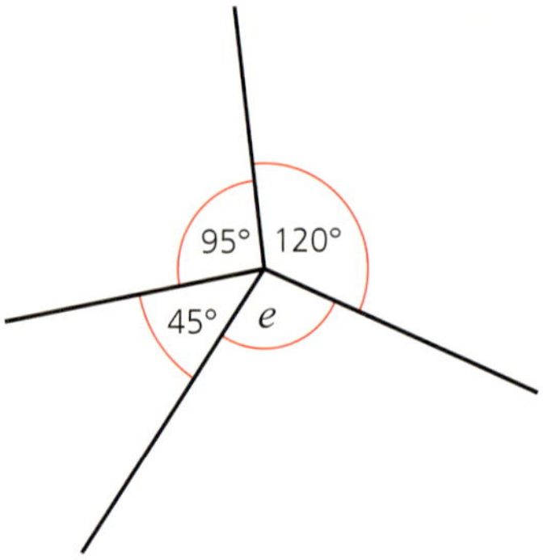

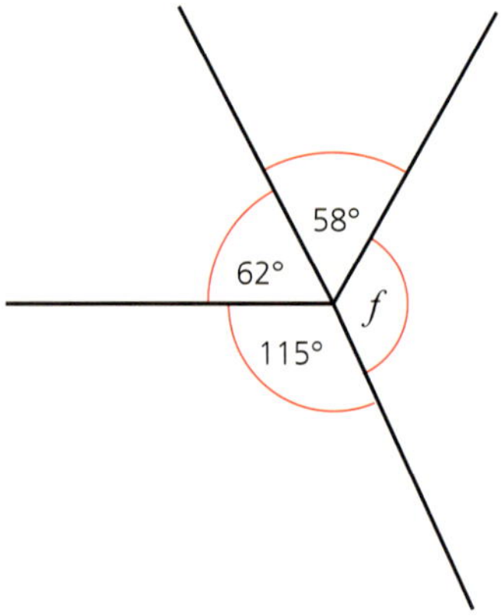

17 Angle properties

2 Find the size of the unknown angles in each of these questions

KEY INFORMATION
If the letter of the angle is the same then the angles must be of equal size.

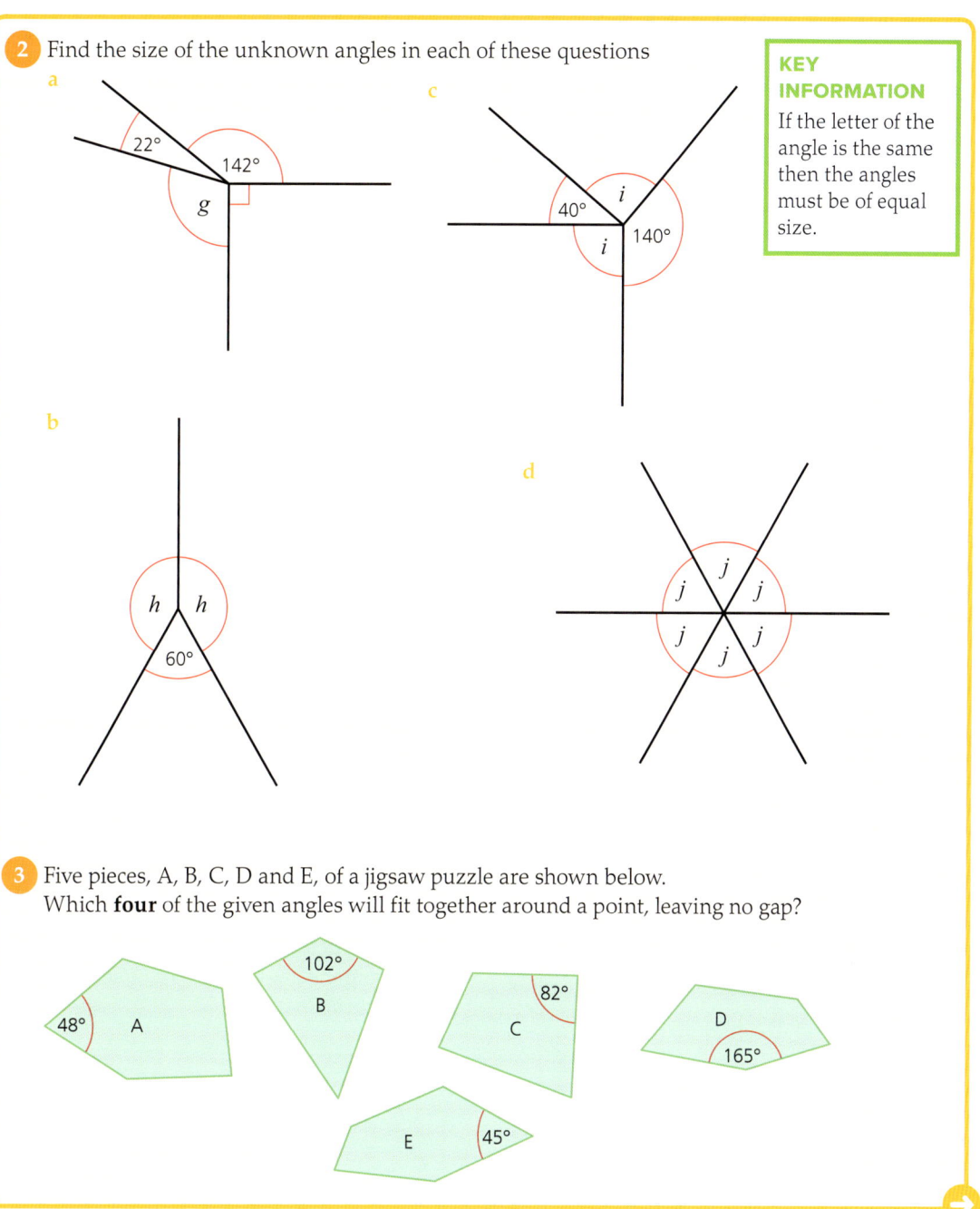

3 Five pieces, A, B, C, D and E, of a jigsaw puzzle are shown below.
Which **four** of the given angles will fit together around a point, leaving no gap?

SECTION 2

 4 Shapes that **tessellate** fit together without leaving any gaps.
For example, **congruent** squares and parallelograms tessellate with themselves as shown:

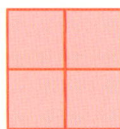 

> **KEY INFORMATION**
> Congruent shapes are exactly the same shape and size as each other.

Does the kite shown below tessellate with other congruent kites? Justify your answer.

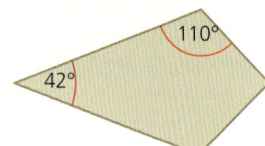

Angles formed within parallel lines

Exercise 17.5

 1 **a** Draw a similar diagram to the one shown below. Measure carefully each of the labelled angles and write them down.

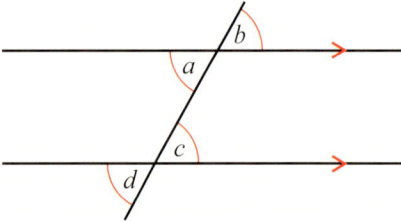

b Draw a similar diagram to the one shown below. Measure carefully each of the labelled angles and write them down.

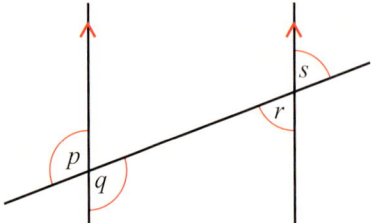

c Draw a similar diagram to the one shown below. Measure carefully each of the labelled angles and write them down.

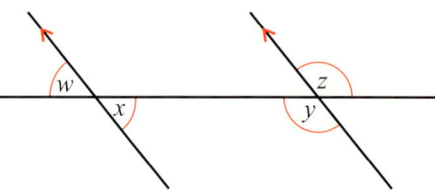

d Write down what you have noticed about the angles you measured in parts (a) – (c).

17 Angle properties

When two straight lines cross, it is found that the angles opposite each other are the same size.

By using the fact that angles at a point on a straight line add up to 180°, it can be shown why they must always be equal in size.

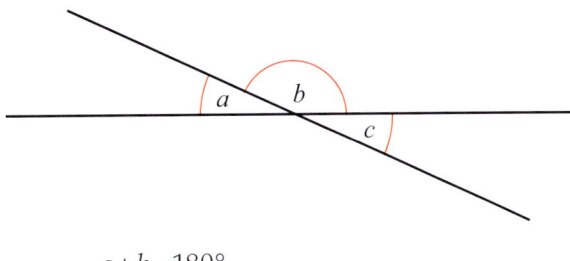

$a + b = 180°$

$c + b = 180°$

Therefore, a is equal to c.

Exercise 17.6

1 a Draw a similar diagram to the one shown below. Measure carefully each of the labelled angles and write them down.

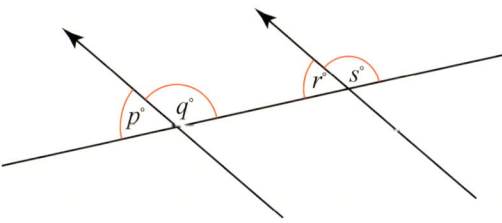

   b Draw a similar diagram to the one shown below. Measure carefully each of the labelled angles and write them down.

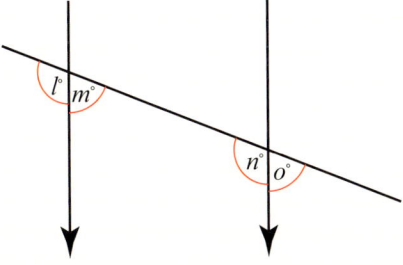

SECTION 2

c Draw a similar diagram to the one shown below. Measure carefully each of the labelled angles and write them down.

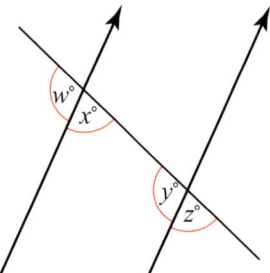

d Write down what you have noticed about the angles you measured in parts (a)–(c).

> **KEY INFORMATION**
> The 'F' formation can be forwards or, as here, backwards.

When a line intersects two parallel lines, as in the diagram below, it is found that certain angles are the same size.

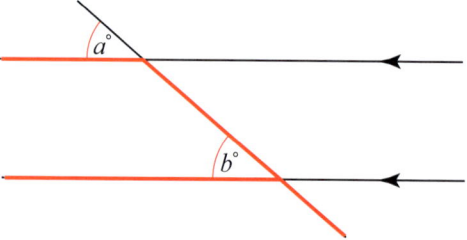

The angles a and b are equal and can be found by looking for an '**F**' formation in a diagram.

A line intersecting two parallel lines also produces another pair of equal angles. These can be shown to be equal by using the two angle relationships found earlier.

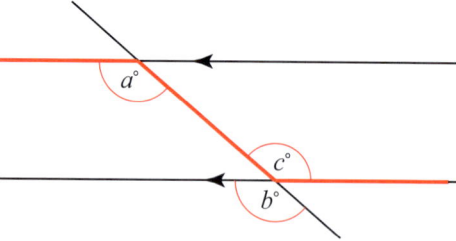

In the diagram above, $a=b$ (found in the F formation), but $b=c$ (as they are opposite). It can therefore be deduced that $a=c$.

These can be found by looking for a '**Z**' formation in a diagram, where the top and bottom lines of the 'Z' are parallel.

17 Angle properties

Exercise 17.7

1 Find the size of each of the unknown angles in these diagrams:

a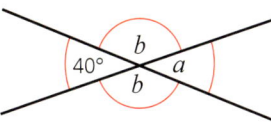

b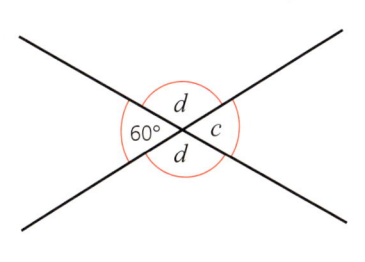

c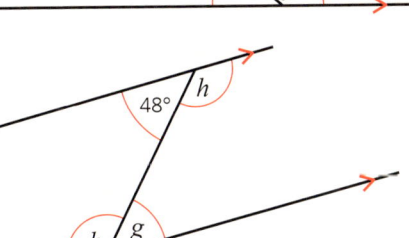

d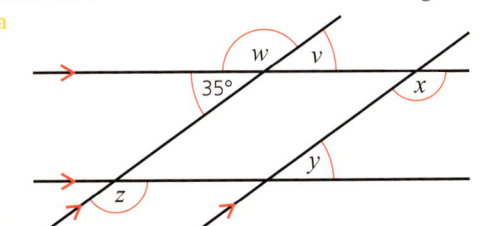

e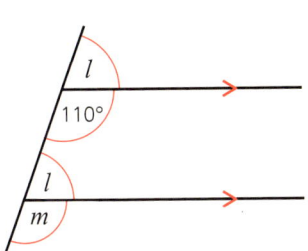

f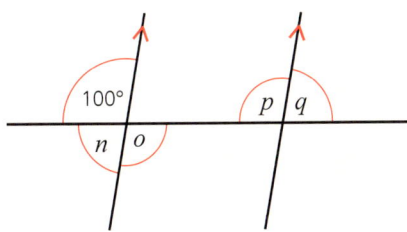

g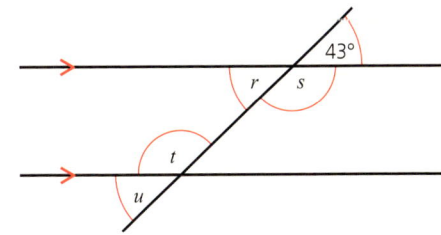

h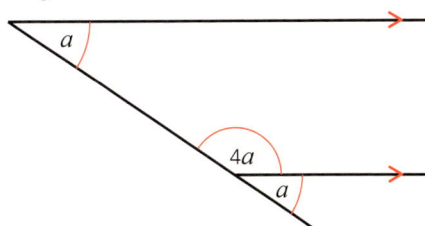

2 Find the size of each of the unknown angles in these diagrams:

a

b

SECTION 2

3 In the diagram below AC and DE are parallel lines.

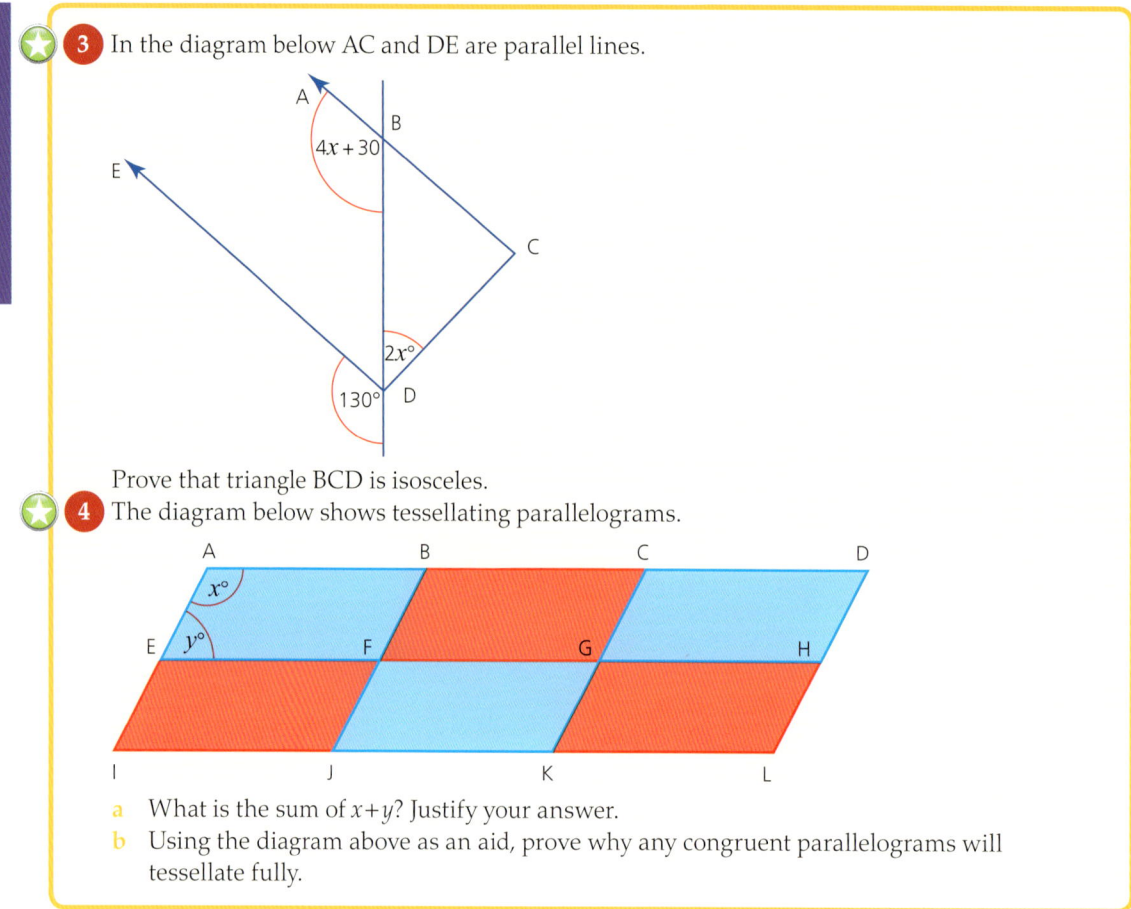

Prove that triangle BCD is isosceles.

4 The diagram below shows tessellating parallelograms.

a What is the sum of $x+y$? Justify your answer.
b Using the diagram above as an aid, prove why any congruent parallelograms will tessellate fully.

Now you have completed Unit 17, you may like to try the Unit 17 online knowledge test if you are using the Boost eBook.

18 Algebraic expressions and formulae

- Understand that a situation can be represented either in words or as an algebraic expression, and move between the two representations.
- Understand that a situation can be represented either in words or as a formula (single operation), and move between the two representations.

Algebraic expressions

You will have seen in Unit 6 that it is possible to represent certain situations as an expression.

For example,

The length of the line is given by the expression $2x+5$.

Being able to write an expression from a given situation is an important mathematical skill to have.

> **Worked examples**
>
> 1 Box A contains x number of sweets.
>   Box B contains ten more than double the number of sweets as box A.
>   Write an expression for the number of sweets in box B.
> $$2x+10$$
>
> 2 A girl is y years old. Her grandmother is four years less than treble the girl's age.
>   Write an expression for the age of the grandmother.
> $$3y-4$$

SECTION 2

Exercise 18.1

1. A student runs a 400 m race in t seconds. His friend manages to run the race four seconds faster. Write an expression, in terms of t, for the time taken by the friend to run the race.

2. A second-hand computer game costs x to buy. A new copy of the game costs six dollars more than double the cost of the second-hand one.
   a Write an expression, in terms of x, of the cost of buying the new copy.
   b If the second-hand computer game costs $5.50, calculate the cost of the new copy.

3. Isabel achieves m marks in a science test. Another student in the class, Luis, gets 15 fewer than triple the amount of marks achieved by Isabel.
   a Write an expression, in terms of m, for the amount of marks achieved by Luis.
   b If Isabel got 22 marks, calculate how many Luis got.

4. Kimie is saving some money in her bank account for a holiday. At the start of the year she has ¥x in her bank account. By the end of the year she has ¥200 less than four times the amount she had at the start of the year.
   a Write an expression, in terms of x, for the amount of money Kimie has in her bank account at the end of the year.
   b If Kimie has ¥19 800 at the end of the year, calculate how much money she had in her bank account at the start of the year.

   > **KEY INFORMATION**
   > ¥ is the symbol for the Japanese currency Yen.

5. Two rectangles A and B are shown below.

   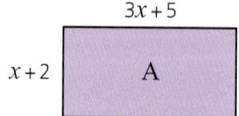
   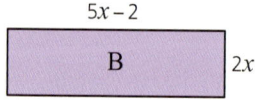

   a If $x = 2$, which rectangle has the greatest perimeter? Justify your answer.
   b If $x = 5$, which rectangle has the greatest perimeter? Justify your answer.

6. Write a mathematical problem to represent the expression $3x - 2$.

7. Write a mathematical problem to represent the expression $6x + 12$.

8 Two beakers of water are shown below.

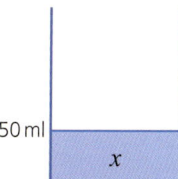 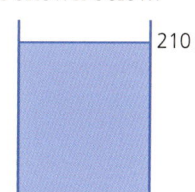

If the amount of water in the first beaker is 50 ml and represented by x, which of the following expressions can represent the amount of water in the second beaker?

$4x + 10$ $3x + 60$ $5x - 40$

Write another expression, in terms of x, to represent the amount of water in the second beaker.

9 There are three jigsaw puzzle boxes.
The first box has n pieces, the second has $3n$ pieces and the third has $2n + 40$ pieces.
  a Is it possible to tell which box has the least number of pieces? Justify your answer.
  b Is it possible to tell which box has the most number of pieces? Justify your answer.

10 Four children are shown below. Their heights are given in terms of x.

Annie: $x - 5$ cm   Billy: x cm   Nehreen: $2x - 60$ cm   David: $x + 10$ cm

  a Name one child who is definitely shorter than another child.
     ……………….. is shorter than…………………
  b Which one child could be either the shortest or the tallest? Justify your answer.

> **LET'S TALK**
> What values of x would make Nehreen the second tallest?

Deriving and using a formula

In Unit 6 you saw that the formula for the perimeter of a rectangle can be derived from a diagram.

It is also possible to derive a formula from written information.

SECTION 2

> **Worked examples**

1. In one year, one dollar ($1) could be exchanged for one euro and twenty cents (€1.20).
   a. Derive a formula to exchange dollars into euros.
      Let d be the number of dollars.
      Let e be the number of euros.
      From the information in the question, $e = 1.2$ when $d = 1$.
      So the formula is:
      $e = 1.2d$
   b. Use your formula to work out how many euros you would receive for $250.
      Substitute $d = 250$ into the formula:
      $e = 1.2 \times 250$
      $e = 300$
      You would receive €300 in exchange for $250.

2. An electrician charges a fee of $25 per hour.
   a. Derive a formula to show his charges, where C is the total charge and n is the number of hours.
      $C = 25n$
   b. What would the electrician charge for working 6 hours?
      $C = 25 \times 6$
      $C = 150$
      The total charge is $150.

3. The distance between two cities is 128 km. The time taken to travel between the two cities is the distance divided by the average speed (s km/h).
   a. Write a formula for the time (T) taken to travel between the two cities.
      $T = \dfrac{128}{s}$
   b. Calculate the time taken to travel between the two cities if the average speed was 80 km/h.
      $T = \dfrac{128}{80} = 1.6$ hours
      The time taken is 1.6 hours.

      > **LET'S TALK**
      > How do you convert 1.6 hours into hours and minutes?

   c. Calculate the average speed if the time taken to travel between the two cities was 2.5 hours.
      Rearranging the formula to make the average speed s the subject gives:
      $s = \dfrac{128}{T}$
      Therefore $s = \dfrac{128}{2.5} = 51.2$.
      The average speed is 51.2 km/h.

18 Algebraic expressions and formulae

Exercise 18.2

1. a Derive a formula for converting the number of hours, h, into minutes, m.
   b Use your formula to convert $4\frac{1}{2}$ hours to minutes.
2. a Derive a formula for converting the number of days, d, into hours, h.
   b Use your formula to convert 14 days to hours.
3. a Derive a formula for converting the number of hours, h, into seconds, s.
   b Use your formula to convert $2\frac{1}{2}$ hours to seconds.
   c Rearrange your formula to convert seconds into hours and use it to work out how many hours are in 27 000 seconds.
4. a Derive a formula for converting the number of metres, m, into centimetres, c.
   b Use your formula to convert 3.3 metres to centimetres.
   c Rearrange your formula to convert centimetres to metres and use it to work out how many metres are in 4550 cm.

▶ Now you have completed Unit 18, you may like to try the Unit 18 online knowledge test if you are using the Boost eBook.

Probability experiments

> You may wish to work in pairs or in groups throughout much of this unit.

- Design and conduct chance experiments or simulations, using small and large numbers of trials. Analyse the frequency of outcomes to calculate experimental probabilities.

Experimental probability

The captain of a volleyball team always calls 'heads' when the coin is tossed to decide which team starts. She lost the toss the first five times. Her friend said, 'Don't worry, you will be right half of the time in the long run.'

'The long run' is sometimes called 'the law of averages' or, by mathematicians, 'the law of large numbers'. It says that, after a very large number of tosses (not just six or seven), the proportion of heads will be close to a half.

The law of averages was first described by the Swiss mathematician Jakob Bernoulli (1655–1704) and is sometimes known as Bernoulli's theorem.

If the coin being tossed is misshapen it would be unfair or biased. The only way to find the probability of heads or tails with that coin, would be by experiment.

This is known as the **relative frequency** and gives the **experimental probability**.

LET'S TALK
Do you think the coin will show head or tails next time? Explain your answer.

What do you think is meant by the friend's statement?

Worked example

a A coin is flipped ten times and it shows heads four times. Based on this result, estimate the experimental probability of getting heads.

Experimental probability is $\frac{4}{10}$ or 0.4.

b The coin is flipped 100 times and it shows heads 42 times. Calculate the experimental probability.

The experimental probability is $\frac{42}{100}$ or 0.42.

c The coin is flipped 1000 times and it shows heads 426 times. Calculate the experimental probability.

The experimental probability is $\frac{426}{1000}$ or 0.426.

d Which of the three values do you think is the most accurate and why?

0.426 is the most accurate as it is based on more results.

19 Probability experiments

Exercise 19.1

1. Take a drawing pin and drop it on the desk 100 times.
   a. Make a tally chart to record the number of drops and the number of times it lands point up.
   b. Use your tally chart to calculate the relative frequency and therefore estimate the experimental probability of the pin landing point up.
   c. Combine your results with nine of your friends, giving a total of 1000 drops. Calculate the experimental probability of the pin landing point up.

2. a. Make a spinner out of card and divide it into four equal parts similar to that shown.
      Stick a pencil through its centre so that it can be spun.
   b. What is the theoretical probability of the spinner landing on each of the colours?
   c. i) Spin your spinner 10 times and record the results in a frequency table.
      ii) How does the experimental probability of your results compare with the theoretical probability?
   d. i) Spin your spinner 100 times and record the results in a frequency table.
      ii) How does the experimental probability of these results compare with the theoretical probability?
   e. Is your spinner biased? Justify your answer.

3. Flip two coins together ten times.
   a. Record the results in a tally chart as HH (two heads), TT (two tails) or HT (a heads and a tails in any order).
   b. Estimate the experimental probability of each.
   c. How do your results compare with theoretical probability of each outcome?
   d. Combine your results with nine friends and estimate the new probabilities.
   e. How do these results compare with theoretical probability?

SECTION 2

4 An Excel spreadsheet can be made to generate random integers between certain limits. To make the spreadsheet **simulate** the rolling of an ordinary dice, generating random integers between 1 and 6 will be needed.

a In cell A1 type '=RANDBETWEEN(1,6)' and then press return. Copy the formula down to cell A12. This will generate 12 random numbers in cells A1 to A12.

> To copy a formula, hold the bottom right corner of the cell and drag it down.

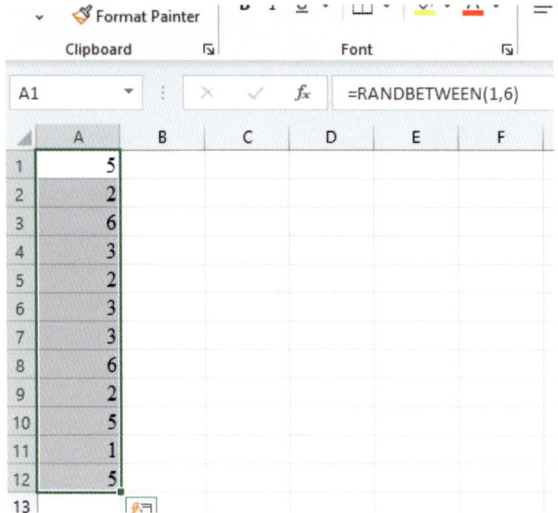

b i) If the random number generator is truly random, what is the theoretical probability of getting each number between 1 and 6?
   ii) How many of each number would you expect to get?
c Draw a tally and frequency table of the results.
d How do the random number generator results compare with the expected results?
e Copy the formula down to cell A1000; this will generate 1000 random numbers in cells A1 to A1000. Using '=COUNTIF(RANGE,VALUE)' count how many times each number appears in the cells and enter your results in a frequency table.

LET'S TALK

For example, typing '=COUNTIF(A1:A1000,1)' will count the number of times the number 1 appears in the cells A1 to A1000. Similarly '=COUNTIF(A1:A1000,2)' will count the number of times the number 2 appears in the cells A1 to A1000 etc.

19 Probability experiments

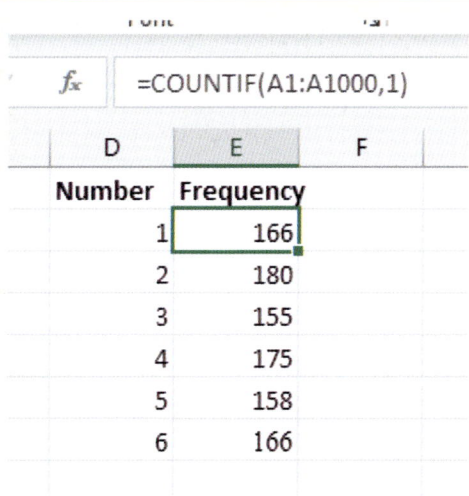

How do these random number generator results compare with the expected results?

f Is the random number generator in Excel truly random? Justify your answer.

5 One student places a total of 50 counters of three different colours in to a bag (for example, they place 30 red, 15 blue and 5 green counters in a bag).

The aim of the activity is for a second student to predict how many counters of each colour are in the bag. The second student takes a counter from the bag, records its colour and then places it back in the bag.

a i) Take a counter from the bag, record its colour in a frequency table and then return it to the bag. Do this ten times.
   ii) From your results can you predict how many of each colour are in the bag?
b i) Repeat the task (a)i) for another 10 counters.
   ii) Did you get the same results as before? Did you expect to get the same results as before? Justify your answer.
   iii) Combining the results from the first 20 counters, can you now predict how many counters of each colour are in the bag?
c i) Repeat the task (a)i) as many times as you think is necessary for you to be able to predict how many counters of each colour are in the bag.
   ii) Make your prediction and then check it against the contents of the bag. How close was your prediction?
d Does repeating an experiment more times, make a probability prediction more accurate? Use your results to justify your answer.

> For question 5 you will need a large supply of coloured counters and to work in pairs.

> Note: You may want to repeat the experiment 100 or even 200 times.

> **LET'S TALK**
> With probability experiments, will you necessarily get the same results each time? Why? Or why not?

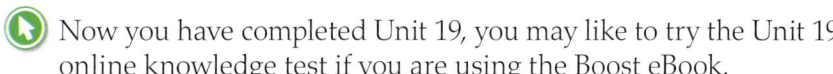

Now you have completed Unit 19, you may like to try the Unit 19 online knowledge test if you are using the Boost eBook.

Introduction to equations and inequalities

- Understand that a situation can be represented either in words or as an equation. Move between the two representations and solve the equation (integer coefficients, unknown on one side).
- Understand that letters can represent an open interval (one term).

Equations

An equation represents two quantities which are equal to each other. To help to see what an equation is, and how it can be used, it is useful to look at it as a pair of scales.

Worked examples

1 Look at these scales.
   The mass of each of the small boxes is 1 kg. Calculate the mass of the large box.

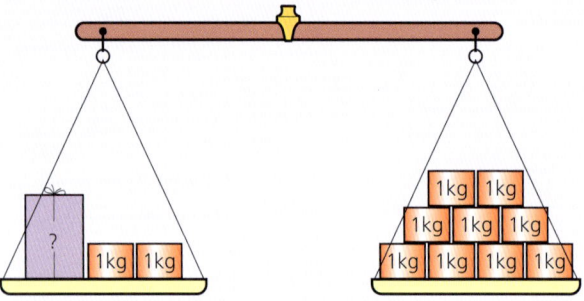

Because the scales are balanced, the left-hand side of the scales must weigh the same as the right-hand side.

In order to keep the scales balanced, whatever we do to the left-hand side of the scales we must also do to the right-hand side.

Take 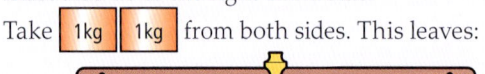 from both sides. This leaves:

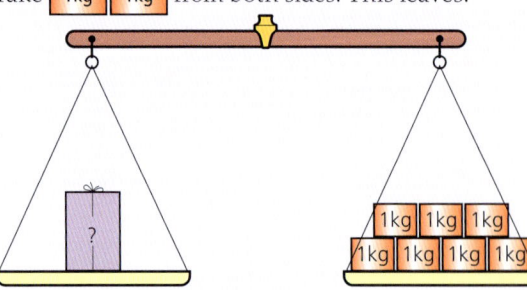

Therefore 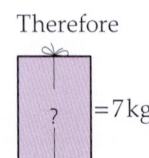 = 7 kg

KEY INFORMATION
By taking the same off both sides, the scales will still be balanced.

162

20 Introduction to equations and inequalities

2 Look at these scales.

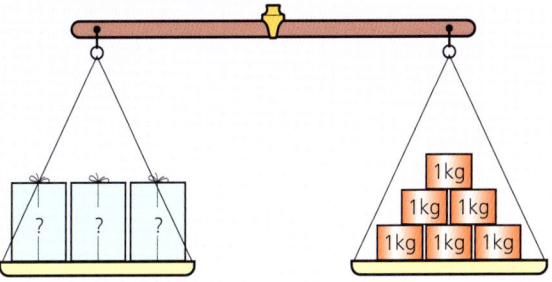

The mass of each of the small boxes is 1 kg. Calculate the mass of each of the large boxes.

We want to find something that we can do to both sides which will leave just one large box on the left-hand side. We cannot take two large boxes from both sides as there are not two to take from the right-hand side. We can divide both sides by 3. This leaves:

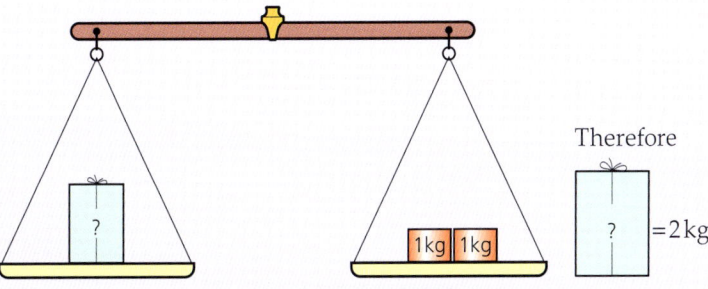

Therefore

? = 2 kg

Algebra

Algebra is used as a simple way of writing equations like the ones in the examples above.

We can use x to represent the mass of a large box.

> We write $1x$ as just x.

Then Worked example 1 becomes:
$$x + 2 = 9$$
$$x + 2 - 2 = 9 - 2 \quad \text{(take 2 kg from each side)}$$
$$x = 7$$

So, the mass of the box is 7 kg.

KEY INFORMATION
Usually the units are not written in the equation, e.g. $3x = 6$ not $3x = 6$ kg.

Worked example 2 becomes:
$$3x = 6$$
$$3x \div 3 = 6 \div 3 \quad \text{(divide both sides by 3)}$$
$$x = 2$$

So, the mass of the box is 2 kg.

SECTION 2

Exercise 20.1

Solve the following equations:

1
- a $a+3=5$
- b $b+4=9$
- c $c+8=15$
- d $d+7=8$
- e $e+12=20$
- f $f+8=11$
- g $g+4=12$
- h $h+9=11$
- i $8+i=10$
- j $4+j=9$
- k $15+k=20$
- l $l+5=6$
- m $17=p+8$
- n $20=q+18$
- o $21=r+4$
- p $15=s+12$
- q $20=t+1$
- r $3u=12$

2
- a $5v=20$
- b $6w=42$
- c $4x=48$
- d $6y=60$
- e $7z=49$
- f $14=7a$
- g $24=8b$
- h $54=9c$
- i $144=12d$

Worked example

Solve the following equation: $3t-4=8$

In this case, in order to get $3t$ on its own on the left-hand side of the equation, we need to add 4 to each side.

$3t-4+4=8+4$
$3t=12$
$t=4$ (divide both sides by 3)

Exercise 20.2

1 Solve the following equations:
- a $2g-3=7$
- b $3h-1=2$
- c $7i-15=6$
- d $3j-18=3$
- e $5k+7=32$
- f $9m+11=74$
- g $6n+12=72$
- h $7r-8=41$
- i $6q-12=84$
- j $3k+7=46$
- k $5m+12=72$
- l $9n+9=72$
- m $6r-8=40$
- n $11q-10=89$

Equations are usually created from real situations and therefore it is important not only to be able to solve an equation but also to be able to form the equation in the first instance.

20 Introduction to equations and inequalities

Worked example

Ayse and Ahmet are playing a 'think of a number' game.

Ayse says, 'I think of a number, double it and subtract 5. The answer is 27.'

What is the number Ayse chose?

Using n to represent the number Ayse chose, we can write:

$2n - 5 = 27$

Then we solve the equation as before:

$2n = 27 + 5$ (add 5 to both sides of the equation)
$2n = 32$
$n = 16$ (divide both sides by 2)

So the number she chose is 16.

LET'S TALK

So far all the equations have just one solution.

Are there examples where there are more than one solution to an equation?

Exercise 20.3

1. Construct an equation from the information given in each question and then solve it.
   a I think of a number and add 10. The answer is 24. What is the number?
   b I think of a number and subtract 7. The answer is 7. What is the number?
   c A number times 7 is 42. What is the number?
   d I think of a number and add 17. The answer is 54. What is the number?
   e I think of a number and subtract 33. The answer is 5. What is the number?
   f A number times 11 is 132. What is the number?

2. A bus has two levels. The ground level can hold a maximum of x passengers. The top floor can hold 16 passengers less than the ground floor.
If the maximum number of passengers the whole bus can take is 112, calculate how many passengers each floor can have. Show your working clearly.

3. The perimeter of the rectangle is 22 cm. Find the length of the longer side. Show your working clearly.

$3x + 1$

4

SECTION 2

 4 Three containers P, Q and R each contain a different number of sweets. Container Q has five more than P. Container R has two less than double the number in P. If the total number of sweets in all three containers is 83, calculate, by forming and solving an equation, how many sweets are in each container.

 5 A rectangle has dimensions as shown.
Explain why if its perimeter is 24 cm its area cannot be 30 cm².

$2x - 1$

7

Inequalities

With equations we are used to the unknown variable having specific values so that one side of the equation is **equal** to the other, for example $4x = 8$. In this case $x = 2$.

With an inequality, however, the variable can take a range of values.

The symbols used with **inequalities** are as follows:

> $>$ means 'is greater than'
> $\geq$ means 'is greater than or equal to'
> $<$ means 'is less than'
> $\leq$ means 'is less than or equal to'

KEY INFORMATION
We can see from this that $x \geq 3$ and $3 \leq x$ mean the same thing.

$x \geq 3$ states that x is greater than or equal to 3, that is, x can be 3, but it could also be 4, 4.2, 5, 5.6 etc.

$3 \leq x$ states that 3 is less than or equal to x, that is, x could still be 3, and could also be 4, 4.2, 5, 5.6 etc.

Therefore:

$5 > x$ can be rewritten as $x < 5$, i.e. x can be 4, 3.2, 3, 2.8, 2, 1 etc.

$-7 \leq x$ can be rewritten as $x \geq -7$, i.e. x can be -7, but also -6.63, -5.1 etc.

These inequalities can also be represented on a number line:

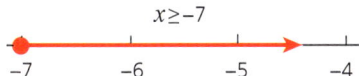

Note that ○→ implies that the number *is not* included in the solution, whilst ●→ implies that the number *is* included in the solution.

> **Worked example**
>
> The maximum number of players (*n*) from one rugby team allowed on the pitch at any one time is 15. Show this information:
>
> a as an inequality
>
> $n \leqslant 15$
>
> b on a number line.
>
>

LET'S TALK

Can *n* be **any** value less than or equal to 15?

KEY INFORMATION

The values that the answer can have will depend on the context of the question, i.e. $n \leqslant 15$ will mean different numbers depending on the question.

Exercise 20.4

1. Represent each of the following inequalities on a number line.
   a $x < 2$ b $x \geqslant 3$ c $x \leqslant -4$ d $x \geqslant -2$

2. Write down the inequalities that correspond to the following number lines:

3. Write the following sentences using inequality signs:
   a The maximum capacity of an athletics stadium is 20 000 people.
   b Five times a number plus 3 is less than 20.
   c The maximum temperature in May was 25°C.
   d A number doubled and then with 6 subtracted from it is greater than 50.

Now you have completed Unit 20, you may like to try the Unit 20 online knowledge test if you are using the Boost eBook.

Section 2 – Review

1. Karl wants to multiply 36 × 24.
   He gets the answer 866.
   Without doing the calculation explain why the answer must be wrong.

2. Four cards are numbered as shown. A fifth card is turned over.

   a What must be the number on the fifth card if:
      i) the median of all five cards is 7
      ii) the mean of all five cards is 9?
   b Explain whether it is possible to work out the number on the fifth card if:
      i) the median of all five cards is 8
      ii) the range of all five cards is 11.

3. The shape below is made from five identical squares.

   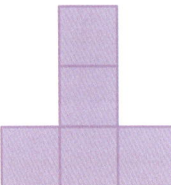

   Copy the diagram and add three more squares to form a shape with a rotational symmetry of order 4.

4. A large rectangle is split into two smaller rectangles. The area of the two rectangles are $6a$ units^2 and 10 units^2 as shown.

   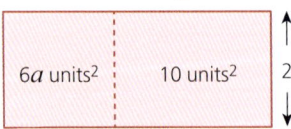

   Write an expression, in its simplest form, for the perimeter of the large rectangle.

Section 2 – Review

5   The pyramid opposite has some numbers already written in some of the blocks.
    The pyramid is constructed in a way such that the numbers in the blocks in the top two rows are the sum of the two fractions in the blocks directly beneath them.
    Calculate the missing fractions in the blocks labelled (a), (b) and (c).

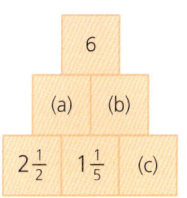

6   A drawing pin, when dropped, can either land point up ↓ or point down ↖.
    Hideko decides to work out the experimental probability of a drawing pin landing point up.
    She drops the pin 1000 times and finds that it lands point *down* 640 times. What is the experimental probability of the pin landing point *up*?

7   Showing your method, calculate the size of angle x.

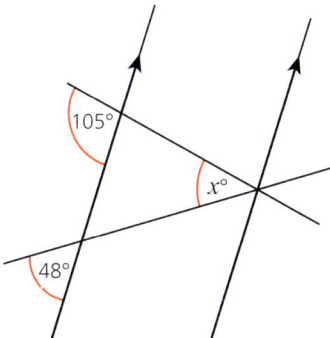

8   The dimensions of two rectangular pieces of card A and B are given below.

Assuming x is positive, are the following statements *always true*, *sometimes true* or *never true*?

a   i) $4x+6$ is bigger than $3x+10$.
    ii) Justify your answer.
b   i) The area of A is smaller than the area of B.
    ii) Justify your answer.

SECTION 2

9 A box contains 80 sweets of three different colours: yellow, red and green.
Two friends decide to try to work out how many of each colour there are without actually counting them. They take a sweet out, record its colour and then put it back in the box.
They do this 360 times and their results are as follows:

	Yellow	Red	Green
Frequency	70	160	130

Estimate how many sweets there are of each colour. Show your working clearly.

10 A kite has dimensions as shown:

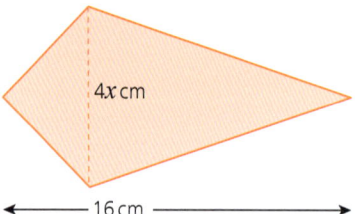

a Explain why the area of the kite is given by the expression $32x$ cm².

b If the area of the kite is 160 cm², calculate the value of x.

SECTION 3

History of mathematics – The development of geometry

The beginnings of geometry can be traced back to around 2000 BCE in ancient Mesopotamia and Egypt. Early geometry was a practical subject concerning lengths, angles, areas and volumes and was used in surveying, construction and astronomy.

The earliest known texts on geometry are the Egyptian Rhind Papyrus (c.1650 BCE), the Moscow Papyrus (c.1890 BCE) and Babylonian clay tablets such as Plimpton 322 (c.1900 BCE).

In the 7th century BCE, the Greek mathematician Thales of Miletus used geometry to solve problems such as calculating the heights of pyramids and the distance of ships from the shore.

Sequences

- Understand term-to-term rules, and generate sequences from numerical and spatial patterns (linear and integers).
- Understand and describe nth term rules algebraically.

Sequences and patterns

A **sequence** is an ordered set of numbers. Each number in the sequence is called a term. The terms of a sequence form a pattern.

Below are examples of three different types of sequences.

2 4 6 8 10 12

In this sequence we are adding 2 to each term in order to produce the next term.

1 2 4 8 16 32

In this sequence we double each term in order to produce the next term.

1 4 9 16 25 36

Here, the difference between consecutive terms increases by 2 each time. It is also the sequence of square numbers.

LET'S TALK
What other well-known sequences of numbers can you think of? Have they got special names?

Sequences in diagrams

Sequences can also be expressed as a series of diagrams. The example below shows the first four diagrams in a sequence of tile patterns.

 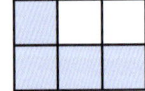 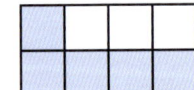 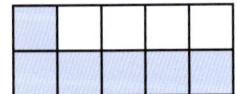

We can see that the tile patterns grow according to a rule. We can enter the numbers of white and blue tiles in each pattern into a table of results.

Number of white tiles	1	2	3	4
Number of blue tiles	3	4	5	6

21 Sequences

> **KEY INFORMATION**
> This is known as a **term-to-term rule**.

> **KEY INFORMATION**
> This is known as a **position-to-term rule**. The position number is the same as the number of white tiles.

There are two types of rules which describe the sequence of blue tiles:
- The number of blue tiles increases by 1 each time.
- The number of blue tiles is always 2 more than the number of white tiles.

The second rule is useful if we know the number of white tiles and want to work out the number of blue tiles. For example, if there are 100 white tiles, how many blue tiles are there?

Number of blue tiles = number of white tiles + 2

Number of blue tiles = 100 + 2 = 102

Exercise 21.1

 1 These diagrams show the first three patterns in a sequence of growing tile patterns.

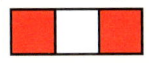 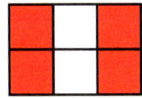 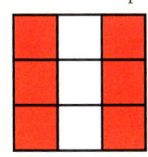

    a Draw the next two diagrams in the sequence.
    b Copy and complete this table.

Number of white tiles	1	2	3	4	5
Number of red tiles					

    c Describe the pattern linking the number of white tiles and the number of red tiles.
    d Use your rule in part (c) to predict the number of red tiles in a pattern with 100 white tiles.

 2

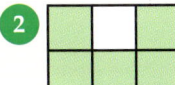 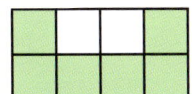 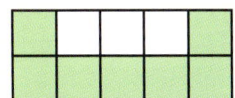

    a Draw the next two diagrams in the sequence.
    b Copy and complete this table.

Number of white tiles	1	2	3	4	5
Number of green tiles					

SECTION 3

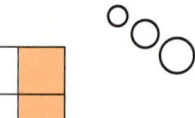

c Describe the pattern linking the number of white tiles and the number of green tiles.
d Use your rule in part (c) to predict the number of green tiles in a pattern with 100 white tiles.

3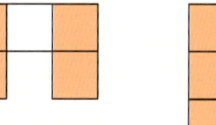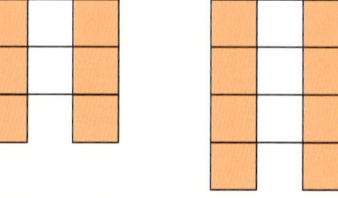

> Look at the diagrams to understand the relationship between the white and green tiles, rather than just looking for a pattern in the table of numbers.

a Draw the next two diagrams in the sequence.
b Copy and complete this table.

Number of white tiles	1	2	3	4	5
Number of orange tiles					

c Describe the pattern linking the number of white tiles and the number of orange tiles.
d Use your rule in part (c) to predict the number of orange tiles in a pattern with 100 white tiles.

4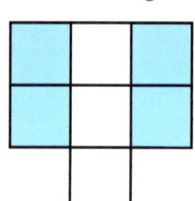

a Draw the next two diagrams in the sequence.
b Copy and complete this table.

Number of white tiles	1	2	3	4	5
Number of blue tiles					

c Describe the pattern linking the number of white tiles and the number of blue tiles.
d Use your rule in part (c) to predict the number of blue tiles in a pattern with 100 white tiles.

Term-to-term rules

A rule which describes how to get from one term to the next is called a term-to-term rule.

21 Sequences

Worked examples

1. Here is a sequence of numbers.

    4   9   14   19   24

    a   Describe the term-to-term rule.

    The term-to-term rule for this sequence is +5.

    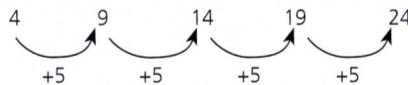

    b   What is the tenth term?

    To calculate the tenth term in the sequence, the pattern can be continued using the term-to-term rule:

    4   9   14   19   24   29   34   39   44   **49**

2. Here is a sequence of numbers.

    1   3   9   27   81

    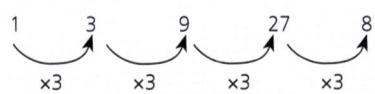

    a   Describe the term-to-term rule for this sequence

    The term-to-term rule for this sequence is ×3.

    b   What is the eighth term in the sequence?

    To calculate the eighth term in the sequence, the pattern can be continued using the term-to-term rule:

    1   3   9   27   81   243   729   **2187**

> **LET'S TALK**
> What is a disadvantage of this type of rule for working out other terms?

Exercise 21.2

1. For each of the following sequences:
    i)   Describe the term-to-term rule.
    ii) Write down the next two terms of the sequence.
    iii) Calculate the tenth term.

    a   2   4   6   8   10
    b   1   3   5   7   9
    c   4   7   10   13   16
    d   2   6   10   14   18
    e   1   8   15   22   29
    f   5   10   20   40   80
    g   2   6   18   54   162
    h   9   7   5   3   1
    i   32   28   24   20   16
    j   144   132   120   108

SECTION 3

The nth term

The method used above for calculating the 10th term of a sequence relies on knowing the term before in order to work out the next one. This method works but can take a long time if the 100th term is needed and only the first five terms are given! A more efficient rule is one which is related to a term's position in a sequence as seen earlier with the tile patterns. This is known as a **position-to-term** rule or the rule for the **nth term**.

Worked examples

1 This table gives the first five terms of a sequence and their positions in the sequence.

Position	1	2	3	4	5
Term	5	6	7	8	9

  a Describe the position-to-term rule.

  By looking at the sequence it can be seen that the term is always the position number $+4$.

  b Write the position-to-term rule as a rule for the nth term.

  The position can be represented by the letter n.

  Therefore, the **nth term** can be given by the expression $n+4$.

Position	1	2	3	4	5	n
Term	5	6	7	8	9	$n+4$

  c Use your rule for the nth term to calculate the 50th term.

  For the 50th term, $n=50$, therefore the 50th term is $50+4=54$.

2 This table gives the first five terms of a sequence and their positions in the sequence.

Position	1	2	3	4	5
Term	4	8	12	16	20

  a Describe the position-to-term rule.

  By looking at the sequence it can be seen that the term is always the position number $\times 4$.

21 Sequences

Remember in algebra $4n$ means the same as $n \times 4$ or $4 \times n$.

b Write the position-to-term rule as a rule for the nth term.
   The position can be represented by the letter n.
   Therefore, the **nth term** can be given by the expression **$4n$**.
c Use your rule for the nth term to calculate the 75th term.
   For the 75th term, $n = 75$, therefore the 75th term is $4 \times 75 = 300$.

Exercise 21.3

1. In the following sequences:
   i) Write down the next two terms.
   ii) Give an expression for the nth term.

   a 6   7   8   9   10
   b 9   10   11   12   13
   c 2   4   6   8   10
   d 8   16   24   32   40
   e −5   −4   −3   −2   −1
   f 100   200   300   400   500
   g −3   −6   −9   −12   −15

2. Look at the tile pattern sequence below.

    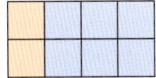  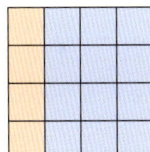

   a Copy and complete the table.

Number of pink tiles	1	2	3	4	5
Number of blue tiles					

   b Describe the relationship between the number of pink and the number of blue tiles.
   c Write an expression for the nth term of the sequence, where n represents the number of pink tiles.
   d If there are 65 pink tiles, how many blue tiles will there be?
   e If there are 540 blue tiles, how many pink tiles are there?

SECTION 3

 3 The sequence of patterns below is made of matchsticks.

Pattern 1   Pattern 2   Pattern 3

   a Write the number of matchsticks in the first five patterns as a sequence of numbers.
   b **Generalise** by writing down the nth term of the sequence, where n is the pattern number.
   c How many matchsticks are there in the 20th pattern?
   d Explain why no pattern will have 232 matchsticks.

4 The grid below shows a pattern of triangles. The first three are shown.
The coordinates of the top vertex of each triangle are given and form a sequence.

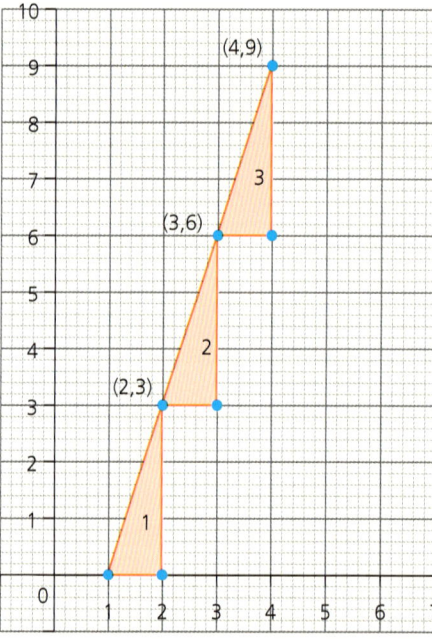

   a What are the coordinates of the top vertex of the 20th triangle? Justify your answer.
   b A triangle in the sequence has a top vertex with coordinates of $(56, y)$.
      i) Which triangle has these coordinates? Justify your answer.
      ii) Calculate the value of y.
   c A triangle in the sequence has a top vertex with coordinates of $(x, 360)$.
Calculate the value of x. Justify your answer.

 Now you have completed Unit 21, you may like to try the Unit 21 online knowledge test if you are using the Boost eBook.

22 Percentages of whole numbers

- Recognise percentages of shapes and whole numbers, including percentages less than 1 or greater than 100.

Percentages of a quantity

Percentages, fractions and decimals are all different ways of representing values. The unique feature of percentages is that they are written as values out of 100. A percentage means 'a part of 100' so 25 per cent, written 25%, means 25 parts in 100.

As a fraction this is written as $\frac{25}{100}$, but it could also be written as $\frac{50}{200}$, $\frac{10}{40}$, $\frac{126}{504}$ or any other equivalent fraction.

Worked examples

1 Find 25% of 60.

   We know that as a simplified fraction 25% is $\frac{1}{4}$. This rectangle has been divided into 60 squares and split into quarters. One of the quarters has been shaded.

   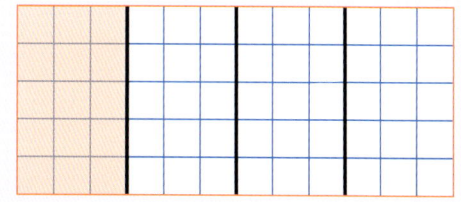

   From the diagram it is clear therefore that 25% of 60 is 15 as 15 squares are shaded.

   Doing a drawing though can be time consuming. Therefore, working it out numerically:

   25% of 60 = $\frac{1}{4}$ of 60 = 60 ÷ 4 = 15.

2 Find 30% of 250.

   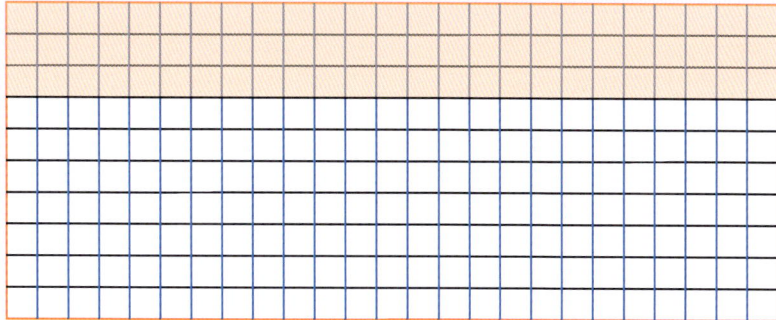

   As a fraction 30% can be written as $\frac{30}{100}$ which when simplified can be written as $\frac{3}{10}$.
   The rectangle above is made up of 250 squares. It has been divided horizontally into tenths. To calculate $\frac{3}{10}$, therefore, three rows have been shaded in.

SECTION 3

$\frac{1}{10}$ of 250 = 25

So $\frac{3}{10}$ of 250 = 3 × 25 = 75

3 Calculate $\frac{1}{2}$% of 600.

$\frac{1}{2}$% means a half out of 100.

Re-writing $\frac{1}{2}$% as a fraction out of 100 therefore gives $\frac{\frac{1}{2}}{100}$ which is equivalent to $\frac{1}{200}$.

Therefore $\frac{1}{2}$% is equivalent to $\frac{1}{200}$.

$\frac{1}{2}$% of 600 = 600 ÷ 200 = 3

KEY INFORMATION

To work out 12.5% of a quantity, compare it to 25%. This will also help when working out 37.5% (25% + 12.5%) and 87.5% (50% + 37.5%).

Exercise 22.1

1 Work out the following percentages.
   a 50% of 300
   b 75% of 400
   c 40% of 600
   d 25% of 360
   e 80% of 250
   f 12.5% of 320
   g 37.5% of 240
   h 87.5% of 880

2 15% of students in a school of 720 students are left-handed. What number of students are right-handed?

3 At a Cambridge college, 65% of the maths students come from England. If there are 240 maths students, how many are not from England?

4 A survey showed that 75% of vehicles passing a school were cars. If 4200 vehicles passed the school in one day, how many were cars?

5 In a village of 3500 people, 12% are left-handed, and 15% of the left-handed people have blonde hair. How many people in the village are blonde and left-handed?

6 a Match each of the percentages on the left to their equivalent fractions on the right.
      Note, that some fractions have no equivalent percentage pair.

   Percentages: 0.5%, 125%, 50%, $\frac{1}{3}$%, 25%, 150%, 30%, 100%, 0.25%

   Fractions: $\frac{5}{4}$, $\frac{3}{20}$, $\frac{1}{200}$, $\frac{1}{2}$, $\frac{1}{300}$, $\frac{1}{8}$, $\frac{3}{30}$, $\frac{1}{4}$, $\frac{3}{2}$, $\frac{1}{400}$, $\frac{3}{10}$, 1

   b For each matched pair in (a) above, calculate that amount out of 600.

180

22 Percentages of whole numbers

 7 a By copying the diagram below twice, use it to show that 20% of 40 is the same as 40% of 20.

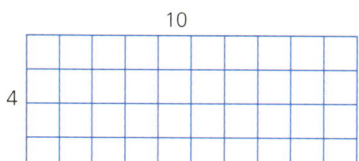

> **LET'S TALK**
> Is $x\%$ of y always the same as $y\%$ of x? Can you prove your answer?

b Copy the diagram below and explain how it can be used to calculate 87.5% of 56.

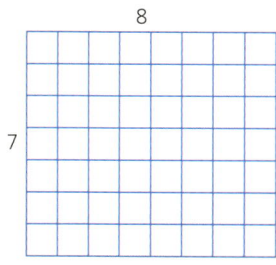

8 A disease affecting cats is known to exist in 0.25% of the cat population. If a vet tests 2800 cats in a year, how many cats would be expected to carry the disease?

9 A small town has a population of 16 200 people. $\frac{1}{3}\%$ of the population are over 100 years old. How many people are over 100?

 10 A 5×4 grid is shown below, with two of its squares shaded.

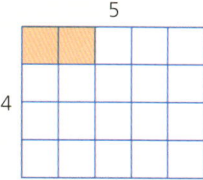

a What percentage of the grid is shaded?
b Four of the grids are arranged together as shown.

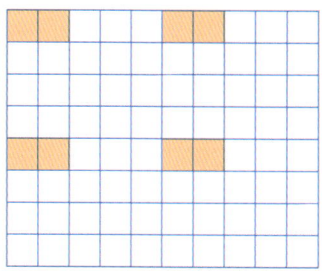

   i) Choosing from the options below, what percentage of the larger grid is now shaded?
      2%   8%   10%   20%   40%
   ii) Justify your answer.

SECTION 3

A quantity as a percentage of another

Percentages, as they are always compared with 100, are a useful way to compare quantities.

Worked examples

1. Three students, Alec, Jaime and Sophia, are comparing results in their tests.
   Alec scored $\frac{32}{40}$, Jaime scored $\frac{55}{70}$ and Sophia scored $\frac{20}{26}$.
   Which student got the highest score?
   As the tests were out of different totals, it is difficult to compare the results as they are.
   A way of comparing them more easily is to convert them to percentages.

   Alec: $\frac{32}{40} \times 100 = 80\%$
   Jaime: $\frac{55}{70} \times 100 = 78.6\%$
   Sophia: $\frac{20}{26} \times 100 = 76.9\%$

   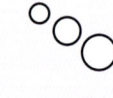

   > To convert a fraction (or a decimal) to a percentage, simply multiply it by 100.

   Therefore Alec got the highest score with 80%.

2. The mass of a puppy increases from 1.25 kg to 10 kg when fully grown. What is the percentage of the adult mass compared with the puppy mass?

   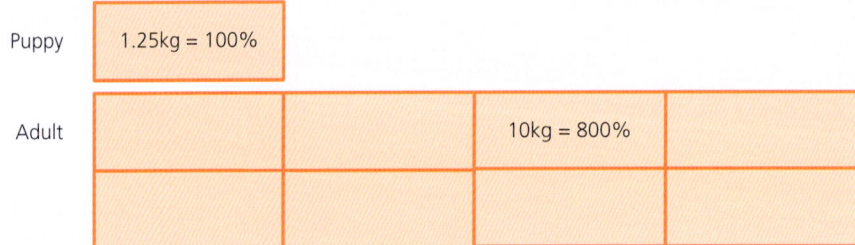

   To work this out it helps to visualise the masses.

   The mass has gone up by a factor of ×8 and therefore the adult mass is 800% of the puppy mass. Note, this is equivalent to a 700% increase not 800%.

   Imagine the puppy's mass as being 100%. The increase in mass from 1.25 kg to 10 kg is equivalent to an increase by a factor of ×8, therefore $8 \times 100 = 800\%$.

   Alternatively, to calculate the adult mass as a percentage of the puppy mass, write the masses as a fraction and multiply by 100.

   $\frac{10}{1.25} \times 100 = 800\%$

22 Percentages of whole numbers

Exercise 22.2

In questions 1–4, write the first quantity as a percentage of the second.

1. a 36 out of 72   b 18 out of 90   c 5 out of 20
2. a 6 out of 60   b 36 out of 60   c 54 out of 60
3. a 100 compared to 80   b 160 compared to 80   c 400 compared to 80
4. a 1 out of 500   b $\frac{1}{2}$ out of 500   c 500 out of 500

5. A school football team plays 25 games in a season. They win 17, draw 2 and lose the rest. Express the numbers won, drawn and lost as percentages of the total number of games played.

6. An airline has a fleet of 80 planes. Of these, 10 are being serviced at any one time. What percentage of the fleet is available for flights?

7. A car manufacturer produces 175 000 sports cars a year. They are only available in white, black and red. 52 500 are white, 78 750 are black and the rest are red.
What percentage of the cars are red?

8. In a class of 32 students in a school, the pupils have the option of playing football (F) and/or tennis (T) for their sports lessons. The numbers are given in the Venn diagram below.

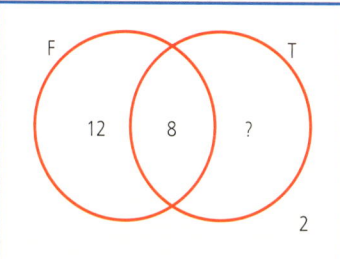

   a What percentage of students play neither sport?
   b What percentage of students play tennis?

9. Yussef takes two maths papers for an exam.
For Paper 1 he scored $\frac{55}{80}$, whilst for Paper 2 he scored $\frac{105}{120}$.
Calculate his percentage score for the whole exam.

10. A farmer sells honey in jars at a market. He has 60 jars to sell.
He calculates that the cost of producing each jar of honey is $1.50.
He manages to sell 25 jars at $3.50 each.
He then decides to reduce the price of the honey to $2.00 for each jar and as a result manages to sell the rest.
   a Calculate how much money he has received from selling his honey as a percentage of cost.
   b Calculate how much profit he has made as a percentage of cost.

SECTION 3

 11 A school has 230 students, all of whom study maths.
It is estimated that on average, each student uses two exercise books per year.
The school can only buy maths exercise books in packs of 25.
Assuming the school buys the minimum number of exercise books necessary for the whole year, what percentage extra does the school buy?

 12 A woman wants to tile her bathroom.
The wall to be tiled is a rectangle with dimensions 2.5 m × 1.2 m.
The tiles she wants measure 10 cm × 10 cm and are only sold in boxes of 90 tiles.
What percentage extra will she have to buy in order to tile her bathroom?

> Note that the dimensions of the tiles are given in cm, whilst the dimensions of the wall are given in metres.

 Now you have completed Unit 22, you may like to try the Unit 22 online knowledge test if you are using the Boost eBook.

Visualising three-dimensional shapes

- Visualise and represent front, side and top view of 3D shapes.

Drawing two-dimensional views of three-dimensional objects

Three-dimensional shapes are often quite difficult to draw in three dimensions. To simplify the drawing process, three-dimensional objects are often drawn in two dimensions.

A common example are architect drawings of houses.

The diagrams below are different two-dimensional views of the same house. By looking at these views it is possible to get all the information about the three-dimensional house.

LET'S TALK
Can you work out how many windows the whole house has from the diagrams?

Front elevation

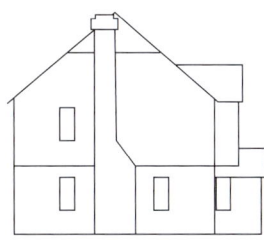
Left elevation

The word 'elevation' simply means 'view'.

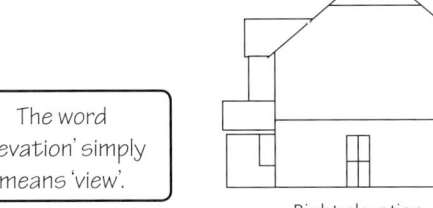
Right elevation

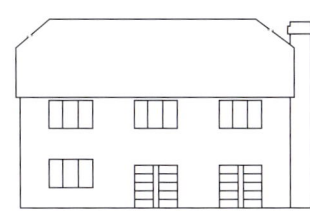
Rear elevation

The four **elevations** shown above all show the sides of the house.

In addition to this, sometimes a view from above is also drawn. A view from above is called a **plan view**.

SECTION 3

KEY INFORMATION

In many cases, you can decide which view is the 'front elevation'. Once that has been decided then the other side views can be labelled.

The plan view is always from above.

KEY INFORMATION

Notice how the faces of the object which are facing the same way have been coloured in with the same colour.

In this example, faces pointing upwards have all been shaded red and those facing the front have been shaded blue.

LET'S TALK

For this 3D object how many 2D views are needed to be able to accurately describe the 3D object? Which views are they?

The diagram below shows a 3D object made from four cubes. Different views are indicated by arrows.

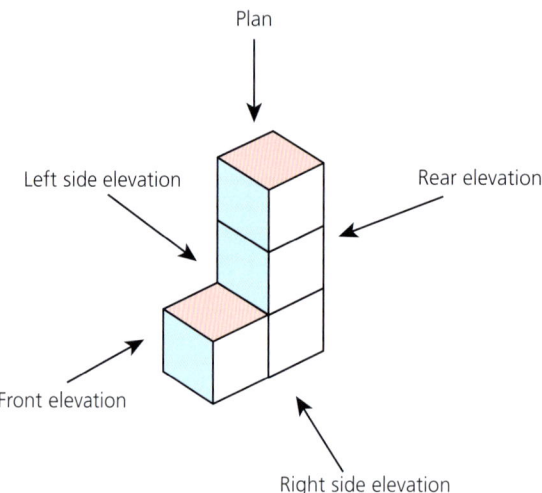

The different views can be drawn in two dimensions as follows:

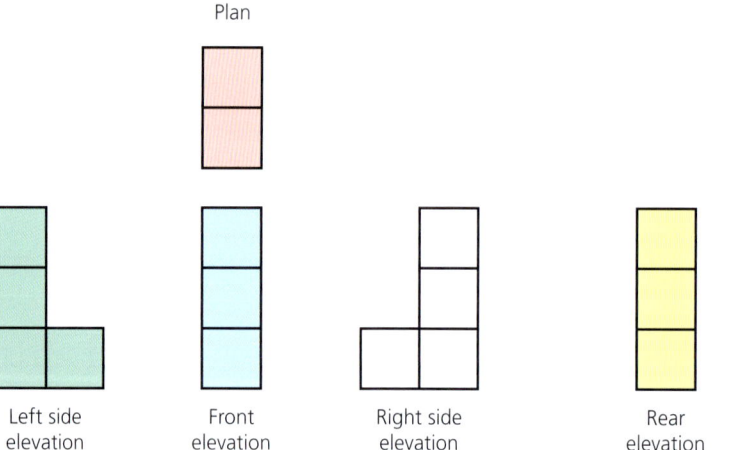

186

23 Visualising three-dimensional shapes

> **Worked example**
>
> The diagram below shows a 3D object made from six cubes.
>
>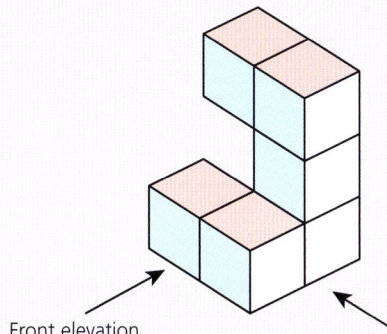
>
> Front elevation
>
> Right side elevation
>
> a Draw 2D drawings of the elevations shown on the diagram.
>
>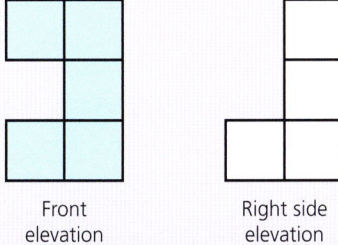
>
> Front elevation   Right side elevation
>
> b Are the two 2D elevations enough to be able to visualise the 3D shape correctly? Justify your answer.
>
> No, the two elevations are not enough.
>
>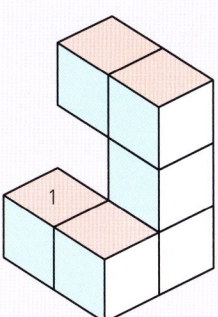
>
> **LET'S TALK**
> What is the minimum number of views needed to accurately represent this 3D object in two dimensions?
>
> If the cube labelled '1' was moved one space to the rear, both the elevations would still look the same.

SECTION 3

Exercise 23.1

In the following exercise, having interlocking cubes will help to see the object from different views. Try **not** to use them for question 5.

 1 a The shape below is made up of five cubes.

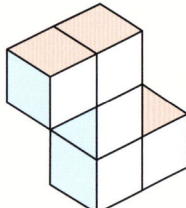

   Explain why some of the faces appear to be triangles.
   b Draw a plan view of the shape.

2 The 3D shape below is made from seven cubes.

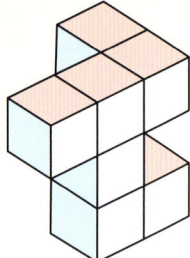

> **LET'S TALK**
>
> Is it possible for this shape to be made from more than seven cubes? What is the maximum number of cubes it could be made from and still look like this from this angle?

A student decides to draw four side elevations (left, right, front and rear) as shown.
   a One of the elevations is incorrect. Which one is it?

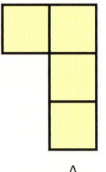 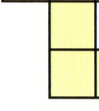 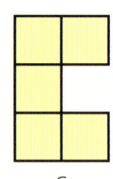 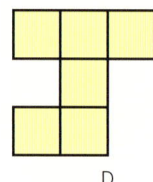

   A   B   C   D

   b Redraw the incorrect elevation correctly.

 3 a The diagram below shows five cubes joined to make a shape.

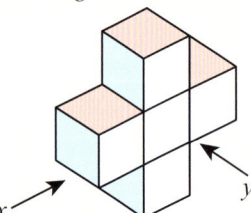

   Draw the two-dimensional elevations seen from directions (*x*) and (*y*).

23 Visualising three-dimensional shapes

b In fact the diagram is made from **six** cubes.
  i) Which of the five visible cubes A, B, C, D and E must the sixth cube be attached to?
  ii) Justify your answer above by drawing a 2D elevation from the (x) direction to show its position.

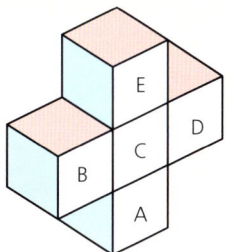

4 The shape below is made from cubes as shown:
  a All the cubes are visible in the 3D diagram.

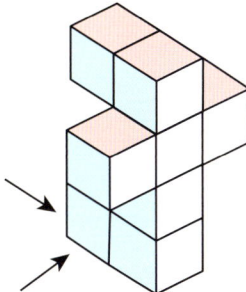

  How many cubes is the shape made from?
  b Draw 2D elevations from each of the directions shown.

5 Two L-shaped objects, each made of four cubes are shown below.
The two objects are then joined so that cube 1 is directly on top of cube 2.

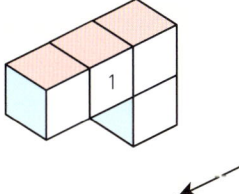

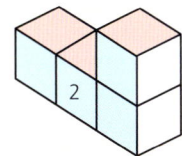

  a Draw a 2D diagram of the view from the direction of the arrow of the joined shape.
  b Draw a plan view of the joined shape.

2D views of common 3D shapes

So far we have only dealt with shapes made from cubes. But you will already know that there are many other three-dimensional shapes.

Examples include a sphere, a cylinder, a pyramid and a cuboid.

189

SECTION 3

Worked example

A square-based pyramid is shown below.

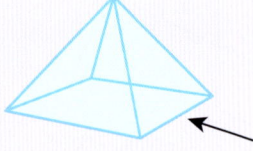

a Draw a side elevation from the direction of the arrow.

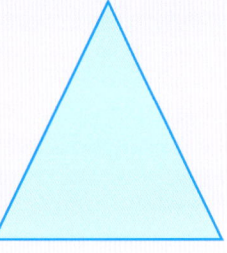

> **KEY INFORMATION**
> Although the face of the pyramid is sloping, in 2D this will look like a flat triangle.

b Draw a plan view.

Exercise 23.2

1. A designer wishes to draw a sphere in two dimensions. What shape will he draw?

2. A cylinder is shown below:

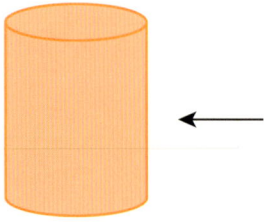

   a Draw a side elevation.
   b Draw a plan view.

23 Visualising three-dimensional shapes

3 A length of guttering to be attached under a roof is shown below. The back of the guttering is taller than the front.

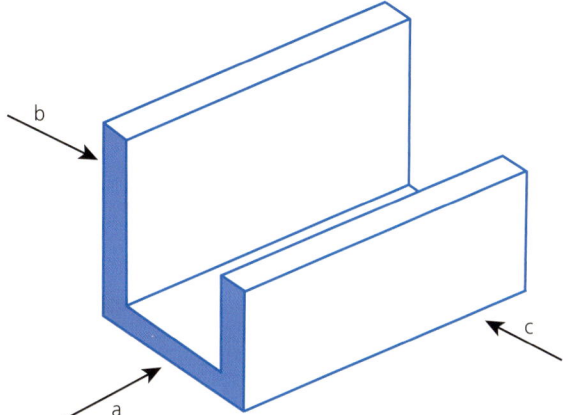

Draw elevations for each of the views labelled a, b and c.

4 The diagram below shows a **truncated cone.**

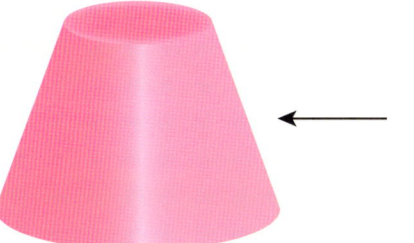

KEY INFORMATION
A cone with the top sliced off is known as a truncated cone.

a Draw the elevation from the direction of the arrow.
b Draw a plan view.

5 Chocolate bars are sometimes sold in packs of 6 as shown.

End face

Chocolates often come in interesting shaped packaging. Research some of the different types and for each, draw their different elevations and plans.

a What shape is the end face of the pack?
b Draw an elevation from the direction of the arrow.
c Will the plan view be different from the elevation drawn in (b) above? Use a diagram to justify your answer.

Now you have completed Unit 23, you may like to try the Unit 23 online knowledge test if you are using the Boost eBook.

Introduction to functions

- Understand that a function is a relationship where each input has a single output. Generate outputs from a given function and identify inputs from a given output by considering inverse operations (linear and integers).
- Understand that a situation can be represented either in words or as a linear function in two variables (of the form $y=x+c$ or $y=mx$) and move between the two representations.

Functions

Modern calculators can work out complex calculations, and can draw graphs and run programs. Early calculators, like this one, were used for simple arithmetic.

You enter the number (the **input**), and the calculator does the arithmetic and produces the answer (the **output**).

A **function machine** works in a similar way to a basic calculator.

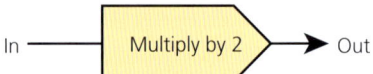

A number is entered. The function machine carries out the mathematical operations according to a rule (the **function**) and produces an output (the answer). In this case, the function is 'multiply by 2'.

Worked examples

1. The numbers 3, 2, 1, 0, −1, −2 are entered into the function machine above.

    Calculate the output in each case.

    The information in the table can also be shown using a diagram known as a **mapping diagram**.

Input	Output
3	6
2	4
1	2
0	0
−1	−2
−2	−4

24 Introduction to functions

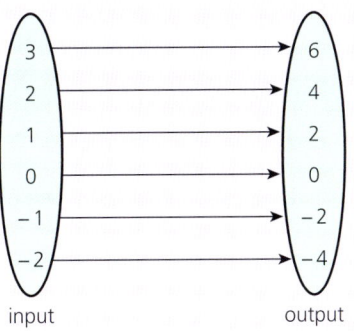

2 A function machine is given as:

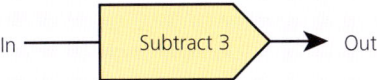

In — Subtract 3 → Out

If the output is 8, calculate the input.

This is done by working out the **inverse function**.

Working backwards, we must work out what mathematical operation undoes the original function.

The opposite of 'Subtraction' is 'Addition'.

The original function can therefore be reversed as follows:

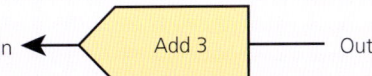

In ← Add 3 — Out

KEY INFORMATION
The inverse function undoes the effect of the original function.

Therefore inserting 8 as the output gives

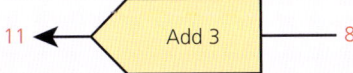

11 ← Add 3 — 8

Therefore if 8 is the output, 11 was the input.

3 Express the equation $y = x + 4$ as a function machine.

Here, whatever value x is given, 4 is added to it to produce the y-value.

Therefore x is the input and y the output.

As a function machine this can be written as

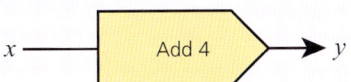

x — Add 4 → y

KEY INFORMATION
The equation $y = x + 4$ can also be called an **algebraic function**.

193

Exercise 24.1

1. Write each of the following functions in machine form:
   a. $y = x+6$
   b. $p = q-5$

In questions 2–4 the input numbers are listed in the table. Calculate the output values.

2.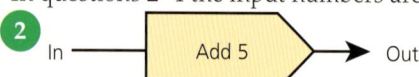

Input	Output
0	
1	
2	
3	

3.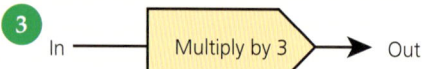

Input	Output
2	
4	
6	
8	

4.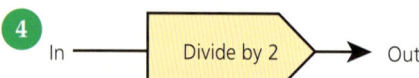

Input	Output
2	
4	
6	
8	

5. In the following question a function machine is given:

   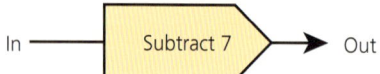

   a. Write down the **inverse** function machine.
   b. Use the function machine and its inverse to complete the following:

Input	Output
5	
	16
15	
	112

6 Four **inverse** function machines are given below:

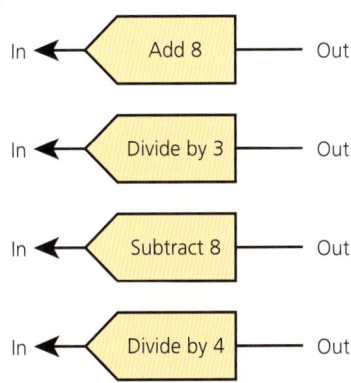

a Which of the four inverse function machines produce the correct input from the given output?

Input	Output
−2	
4	12
8	
20	

b Complete an input/output table for each of the inverse function machines answered in part (a).

 7 Below are two function machines **P** and **Q**.

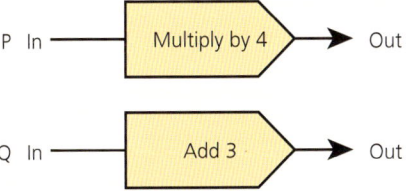

a If both function machines have the same input, calculate the value of that input if the output of **P** is twice the output of **Q**.

b If both function machines have the same output, calculate the value of that output if the input of **Q** is three times the input of **P**.

SECTION 3

Constructing function machines

As in Unit 20 with equations, function machines (and algebraic functions) can be constructed from real situations. In the real world this is important to be able to do, as it will enable you to work out **outputs** from a variety of **inputs**.

> **Worked example**
>
>
>
> Write, in function machine form, the relationship between the number of elephants (e) and the number of elephants' legs (l).
>
> a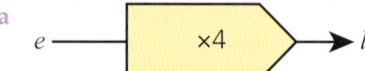
>
> b Write your function machine as an algebraic function.
> $l = 4e$
>
> c If there are 68 elephant legs, how many elephants are there?
> So the inverse function machine needed to answer this question is
>
>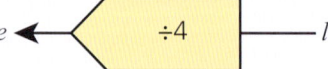
>
> Therefore the number of elephants is $68 \div 4 = 17$.

Assume that all the elephants have the correct number of legs!

24 Introduction to functions

Exercise 24.2

1. A grandmother is 56 years older than her grandson.
   Let the grandmother's age be M and the grandson's age n.
   Write an algebraic function to work out the grandson's age from the grandmother's age.

2. A plumber charges $15 per hour of work.
   a Write a function machine to work out the amount he earns p for working h hours.
   b On one job he works 22 hours. Calculate how much he earns for that work.

3. Claudia lives in Brazil and is going on holiday to India.
   She goes to the bank to change Brazilian currency **Real** (R) to Indian currency **Rupee** (r). The exchange rate is $1R = 17.4r$.
   a Write a function machine to convert Reals to Rupees.
   b If she changes 500 Reals, how many Rupees will she receive?
   c At the end of her holiday, Claudia exchanges 250 Rupees back to Reals. How many Reals will she receive?

LET'S TALK
In real life, why is the exchange rate for changing currency B to currency A, not the **inverse** of the exchange rate of changing currency A to B?

4. A taxi firm charges 80 cents for each kilometre travelled.
   a Which of the following function machines represents the cost $C of travelling n kilometres?

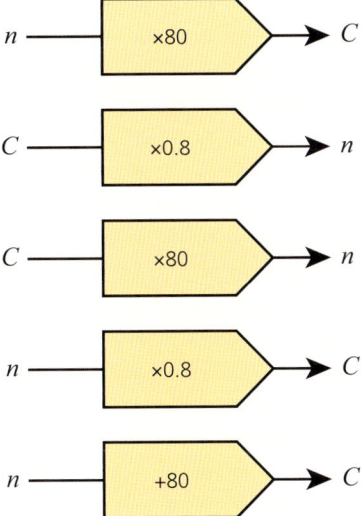

   b Explain why **two** other function machines are incorrect.
   c Use the correct function machine to calculate the cost of taking the taxi for a 55 km journey.

Now you have completed Unit 24, you may like to try the Unit 24 online knowledge test if you are using the Boost eBook.

25 Coordinates and two-dimensional shapes

- Use knowledge of 2D shapes and coordinates to find the distance between two coordinates that have the same x or y coordinate (without the aid of a grid).
- Use knowledge of translation of 2D shapes to identify the corresponding points between the original and the translated image, without the use of a grid.

Coordinates and shapes

Recap

The position of a point in two dimensions (2D) is given in relation to a point called the **origin**. We draw two axes at right angles to each other. The horizontal axis is the *x*-axis, whilst the vertical axis is called the *y*-axis.

The position of point A is given by two coordinates, the *x*-coordinate and the *y*-coordinate. So the coordinates of point A are (3, 2). Similarly, the coordinates of point B are (2, 4).

The axes can be extended in both directions. By extending the *x*- and *y*-axes below zero this grid is produced.

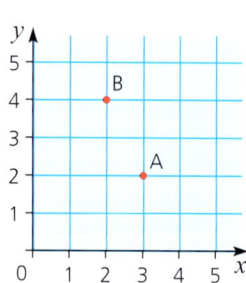

We can describe points C, D and E by their coordinates.

Point C is at (3, −3).

Point D is at (−4, −3).

Point E is at (−4, 3).

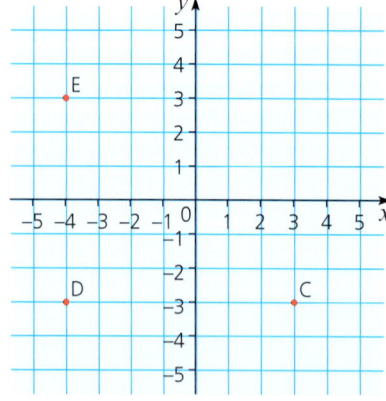

25 Coordinates and two-dimensional shapes

Exercise 25.1

1 Draw a grid with centre $(0,0)$, the origin, and mark the x- and y-axes with scales from -8 to $+8$. Mark these points on your grid.
- a A(5, 2)
- b B(7, 3)
- c C(2, 4)
- d D(−8, 5)
- e E(−6, −8)
- f F(3, −7)
- g G(7, −3)
- h H(6, −6)

For each of questions 2–4, draw a separate grid with x- and y-axes from -6 to $+6$. Plot the points, join them up in order and name the shape you have drawn.

2 A(3, 2)   B(3, −4)   C(−2, −4)   D(−2, 2)

3 D(1, 3)   E(4, −5)   F(−2, −5)

4 G(−6, 4)   H(0, −4)   I(4, −2)   J(−2, 6)

In the questions above, the coordinates of all the **vertices** for each shape were given.

However, as you will recall from your work in Unit 2, many shapes have particular properties. These properties can help us deduce the coordinates of missing vertices.

Worked example

The coordinates of three vertices of a rhombus are given in the diagram.

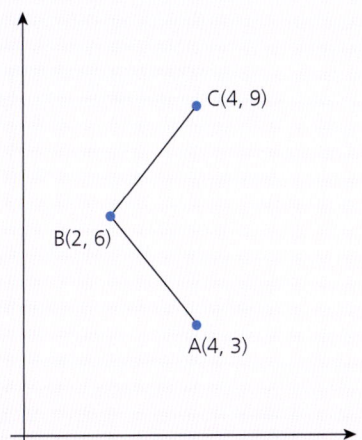

LET'S TALK
Here three vertices of a rhombus are given in order to deduce the fourth vertex. Can all the vertices of a rhombus be deduced if only two are given? What about other quadrilaterals?

a Deduce the coordinates of the missing vertex D.
The coordinates of D are (6, 6).

SECTION 3

b Justify your answer using the properties of the rhombus.

The diagonals of a rhombus intersect at right angles.
Vertices A and C lie in the same vertical line, because their x-coordinates are the same.
Therefore, D must be on the same horizontal line as B.
A rhombus also has two lines of symmetry, therefore D must have coordinates (6, 6) as shown.

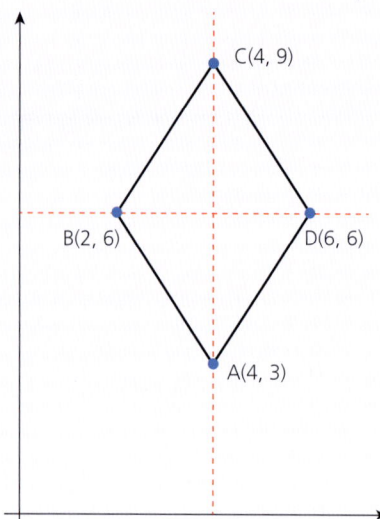

Notice how the numbers on the axes do not need to be given in order to work out the missing vertex. This is because the properties of the rhombus mean that the 4th vertex can only be at (6, 6).

Exercise 25.2

1 The axes below show three vertices of a square ABCD.

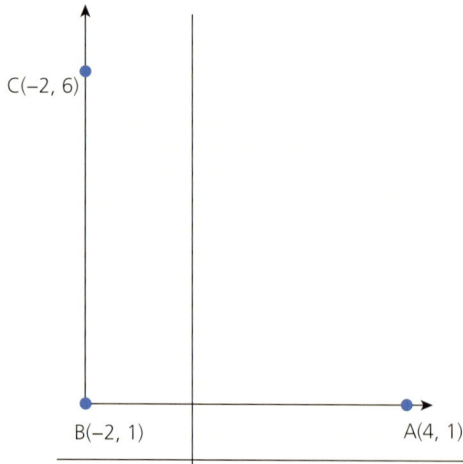

LET'S TALK
If instead of being a square, ABCD was a kite, what could the possible coordinates of vertex D be? What if it was an arrow head?

a What are the coordinates of vertex D?
b Justify your answer by referring to the distance between the vertices.

25 Coordinates and two-dimensional shapes

2 Three vertices of parallelogram WXYZ are shown in the axes below.

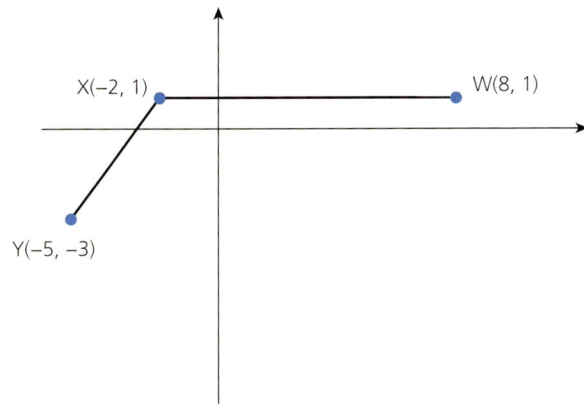

   a A student states that the *y* coordinate of vertex Z is −5. Explain why this must be wrong.
   b Deduce the coordinates of Z.

3 a Draw a grid with *x*- and *y*-axes from −10 to +10.
      Plot the points P(−6, 4), Q(6, 4) and R(6, −2).
      i)   Plot point S such that PQRS is a rectangle.
      ii) Write down the coordinates of S.
      iii) Draw diagonals PR and QS. What are the coordinates of their point of intersection?
      iv) What is the area of PQRS?
   b i)   On the same grid plot points M(−8, 4) and N(4, 4).
      ii) Join points MNRS. What shape have you drawn?
      iii) What is the area of MNRS?
      iv) Justify your answer to (b) iii).
   c i)   On the same grid plot point J where point J has *y*-coordinate +10 and JRS forms an isosceles triangle.
      ii) Write down the coordinates of J.
      iii) What is the *x*-coordinate of all the points on the line of symmetry of triangle JRS?

4 One side of a square PQRS is given on the axes below.

a   If the diagonals of the square intersect at the origin, calculate the coordinates of the two missing vertices R and S.
b   If the diagonals of the square intersect at (3, −3) what are the coordinates of the vertices R and S?

5   Three vertices of a kite ABCD are shown below.

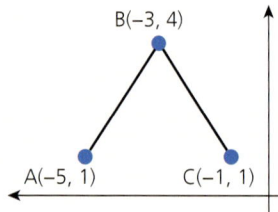

a   What must the x-coordinate of vertex D be? Justify your answer
b   The area of the kite ABCD is 20 units². Deduce the coordinates of vertex D.

6   One side of a rectangle LMNO is given in the diagram below. The coordinates of the centre of the rectangle C are (−1, −3.5).

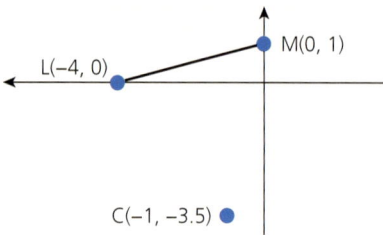

Calculate the coordinates of the two missing vertices N and O.

Coordinates and translation of 2D shapes

You will already be familiar with **translation** and how to describe that movement.

When a shape is translated, every point on the shape is translated by the same amount.

KEY INFORMATION
Translation is a sliding movement. To describe a translation you need to give how far the shape has moved horizontally and how far it has moved vertically.

25 Coordinates and two-dimensional shapes

Worked example

A triangle ABC is shown on the axes below.

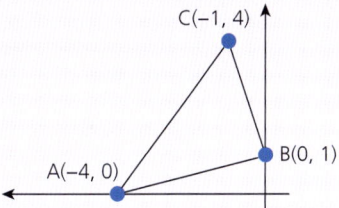

The triangle is translated and the vertex A moves to the new coordinates (1, −2).
  a Describe the translation.
     Vertex A has been translated 5 units to the right and 2 units down to its new position.
  b Calculate the coordinates of the vertices B and C after the translation.
     All points move the same amount, i.e. 5 units to the right and 2 units down.
     Therefore the coordinates of B are (5, −1) and those of C are (4, 2).

Exercise 25.3

1 The axes below shows triangle ABC and its position to A'B'C' after a translation.

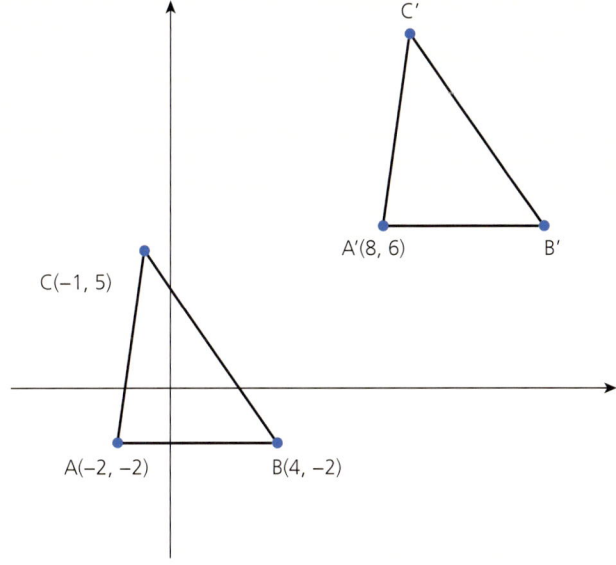

SECTION 3

a   Describe the translation that maps ABC on to A'B'C'.
b   Deduce the coordinates of B'.
c   Deduce the coordinates of C'.
d   Explain why the area of triangle A'B'C' is 21 units².

2   A square PQRS is shown on the grid below.

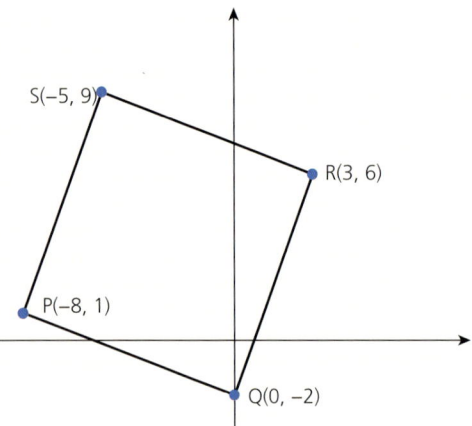

LET'S TALK
How could you calculate the area of this square?

The square is translated to a new position P'Q'R'S'.
a   One of the following coordinates of the translated square P'Q'R'S' is incorrect. Which one?
    P'(−4 , 5)   Q'(4 , 3)   R'(7 , 10)   S'(−1 , 13)
b   Justify your answer to (a).

3   The diagonals of the rectangle EFGH intersect at a point M.
The rectangle EFGH is translated to a new position E'F'G'H' and the intersection point of the diagonals is at the origin.

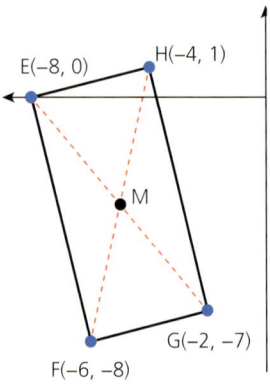

KEY INFORMATION
The intersection point of the diagonals is halfway between the opposite vertices.

a   Calculate the coordinates of the vertices E', F', G' and H'.
b   Justify your answers in part (a).

 Now you have completed Unit 25, you may like to try the Unit 25 online knowledge test if you are using the Boost eBook.

26 Squares, square roots, cubes and cube roots

- Understand the relationship between squares and corresponding square roots, and cubes and corresponding cube roots.

Squares

This pattern sequence is made up of 1 cm squares.

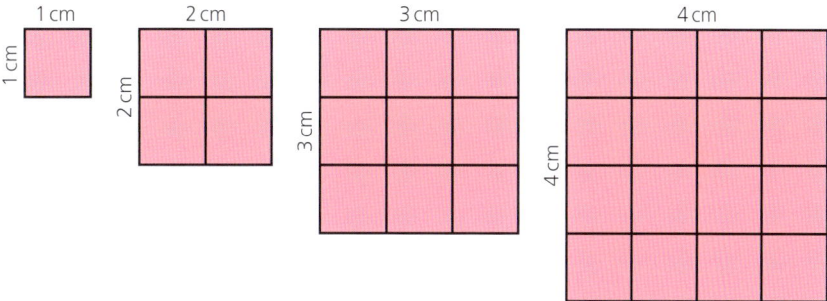

The 1 cm × 1 cm square contains one 1 cm × 1 cm square.

The 2 cm × 2 cm square contains four 1 cm × 1 cm squares.

The 3 cm × 3 cm square contains nine 1 cm × 1 cm squares.

The 4 cm × 4 cm square contains sixteen 1 cm × 1 cm squares.

The numbers 1, 4, 9, 16 are square numbers, and are made by multiplying an integer (whole number) by itself. For example,

   8×8=64

Therefore 64 is a **square number**.

But

   2.3×2.3=5.29

5.29 is not a square number as 2.3 is not an integer.

Squaring a number is multiplying a number by itself. For example,

   8 squared is 8×8

   2.3 squared is 2.3×2.3

SECTION 3

KEY INFORMATION
A number raised to the power of 2 is also that number squared.

There is a short way to write a number squared. It involves using indices. For example,

$8 \times 8 = 8^2$

$2.3 \times 2.3 = 2.3^2$

Using a calculator

The squared button on the calculator usually looks like this .

Worked example

Use the calculator to evaluate 17^2.

LET'S TALK
If you had got an answer of 256, how could you tell immediately that the answer must be incorrect?

Square roots

The inverse (opposite) operation to addition is subtraction, the inverse operation to multiplication is division. Squaring also has an inverse operation. It is known as the **square root**.

Using a calculator

All scientific calculators can work out the square root of a number by using the key.

KEY INFORMATION
These instructions are for calculators that use direct algebraic logic. For some calculators, particularly older ones, the calculation needs to be entered differently. Check how yours works.

Worked example

Use a calculator to work out $\sqrt{729}$.

26 Squares, square roots, cubes and cube roots

Calculations without a calculator

Worked examples

1. Without using a calculator evaluate $\sqrt{0.36}$.

   0.36 can be written as a fraction.
   $0.36 = \frac{36}{100}$
   $\sqrt{0.36} = \frac{\sqrt{36}}{\sqrt{100}} = \frac{6}{10}$
   $\frac{6}{10} = 0.6$
   Therefore $\sqrt{0.36} = 0.6$.

2. Without using a calculator evaluate $\sqrt{0.81}$.

   0.81 can be written as a fraction.
   $0.81 = \frac{81}{100}$
   $\sqrt{0.81} = \frac{\sqrt{81}}{\sqrt{100}} = \frac{9}{10}$
   $\frac{9}{10} = 0.9$
   Therefore $\sqrt{0.81} = 0.9$.

Exercise 26.1

1.
   a. Evaluate the following without using a calculator.
      i) $\sqrt{25}$　　ii) $\sqrt{9}$　　iii) $\sqrt{121}$　　iv) $\sqrt{169}$
   b. Check your answers to (a) using a calculator.

2. A child is covering a square board with mosaic tiles.

   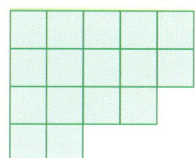

   Not to scale.

   The board is 40×40 cm.
   Each mosaic tile is 2×2 cm.
   a. How many mosaics are needed for one complete row?
   b. The mosaics come in packs of 30.
      i) How many packs will the child need?
      ii) How mosaics will be left over?

SECTION 3

3 Evaluate the following without using a calculator.

a) $\sqrt{\frac{1}{9}}$ d) $\sqrt{\frac{9}{100}}$ g) $\sqrt{0.01}$

b) $\sqrt{\frac{1}{49}}$ e) $\sqrt{\frac{25}{36}}$ h) $\sqrt{0.09}$

c) $\sqrt{\frac{4}{9}}$ f) $\sqrt{\frac{49}{81}}$

LET'S TALK

Can you think of two other square numbers which add together to make another square number?

How many can you find in 10 minutes?

4 A tiler has 100 square tiles for tiling two square panels in a bathroom.

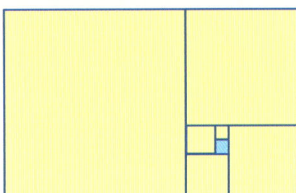

100 tiles

a Explain why the panels cannot be of the same size.
b Each tile is 10×10 cm. What are the dimensions of the two square panels if all 100 tiles are used?

5 The pattern below shows an arrangement of squares.

LET'S TALK

Research the mathematician Fibonacci and in particular the Fibonacci sequence.

How does this pattern relate to the Fibonacci sequence?

The small blue square has dimensions 1×1 units
a How many blue squares would fit into the whole of the pattern?
b Use your answer to (a) to explain why the whole shape cannot itself be a square.

Cubes

This pattern sequence is made up of 1 cm cubes.

26 Squares, square roots, cubes and cube roots

The 1 cm × 1 cm × 1 cm cube contains **one** 1 cm × 1 cm × 1 cm cube.

The 2 cm × 2 cm × 2 cm cube contains **eight** 1 cm × 1 cm × 1 cm cubes.

The 3 cm × 3 cm × 3 cm cube contains **27** 1 cm × 1 cm × 1 cm cubes.

The numbers 1, 8 and 27 are **cube numbers**, and are made by multiplying an integer by itself and then by itself again. For example,

$5 \times 5 \times 5 = 125$

Therefore 125 is a cube number.

Cubing a number is multiplying a number by itself three times. As with squaring, there is a short way to write a number cubed using indices. For example,

$5 \times 5 \times 5 = 5^3$

Using a calculator

Not many calculators have a specific $\boxed{x^3}$ key. However, all scientific calculators have an 'indices' key. The indices key looks like this: $\boxed{y^x}$. It allows a number to be raised to any power, not just cubed.

Worked example

Use a calculator to work out 8^3.

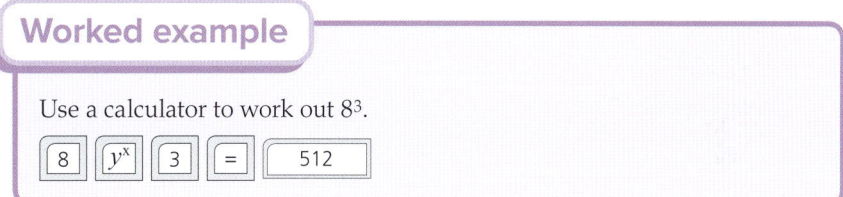

Cube roots

The inverse of cubing a number is finding its **cube root**, written as $\sqrt[3]{}$.

So $\sqrt[3]{125}$ is 5 (since $5 \times 5 \times 5 = 125$) and $\sqrt[3]{343}$ is 7 (since $7 \times 7 \times 7 = 343$).

SECTION 3

Exercise 26.2

1. How many 1 cm × 1 cm × 1 cm cubes would make up cubes with the following side lengths?
   a 4 cm
   b 6 cm
   c 10 cm
   d 9 cm

2. Using the y^x button on your calculator, evaluate the following.
   a 11^3
   b 20^3
   c 2.5^3
   d 6.2^3

3. A sugar cube manufacturer wants to sell 500 sugar cubes in a box also in the shape of a cube.
   a Assuming there are no gaps in the box, explain why this will not be possible.
   b What is the closest number of cubes to 500 that he will be able to pack into a cube shaped box? Give a **convincing** reason for your answer.

4. Without using a calculator, work out the cube roots of the following numbers.
   a 8
   b 125
   c 27
   d 1000

5. Without using a calculator if possible, work out the following.
   a $\sqrt[3]{64}$
   b $\sqrt[3]{343}$
   c $\sqrt[3]{216}$
   d $\sqrt[3]{800}$

6. A shop owner buys a three-digit combination number padlock for his door.
   He wants to choose a number which he can remember.

   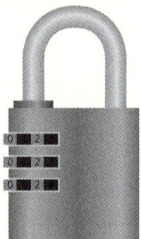

   > **LET'S TALK**
   >
   > Is there a three-digit number he could choose that was both a square number and a cube number?

   He decides to choose a two-digit number (i.e. a three-digit number starting with zero) which is both a square number and a cube number. Which combination does he choose?

7. A child's box of building bricks contains the following numbers of cubes:
   Eight cubes with dimensions 2 × 2 × 2 cm.
   Sixty cubes with dimensions 1 × 1 × 1 cm.
   Petra wants to build a large 5 × 5 × 5 cm cube with the cubes in her box.
   Has she got enough cubes to build it? Justify your answer.

Now you have completed Unit 26, you may like to try the Unit 26 online knowledge test if you are using the Boost eBook.

27 Linear functions

- Use knowledge of coordinate pairs to construct tables of values and plot the graphs of linear functions, where y is given explicitly in terms of x.
- Recognise straight-line graphs parallel to the x- or y-axis.

Graphs of linear functions

A line is made up of an infinite number of points. The coordinates of every point on a straight line all have a common relationship.

In other words, the x- and y-values follow a pattern. It is this pattern that gives the equation of the line.

The line below is plotted on a pair of axes.

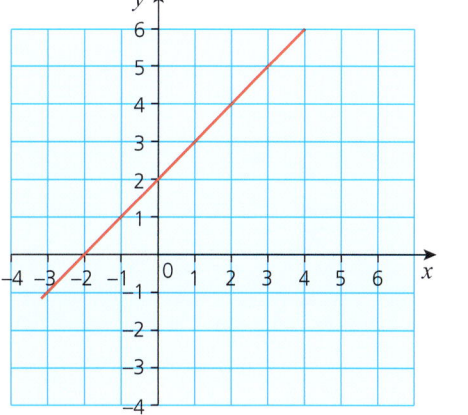

x	y
0	2
1	3
2	4
3	5
4	6

By looking at the coordinates of some of the points on the line we can see a pattern.

In words, the pattern linking the x- and y-coordinates can be described as follows:
- The y-coordinates are 2 more than the x-coordinates.
- Written algebraically, this is $y = x + 2$.

This is known as the **equation of the straight line**.

SECTION 3

Worked examples

1. Look at the graph below.

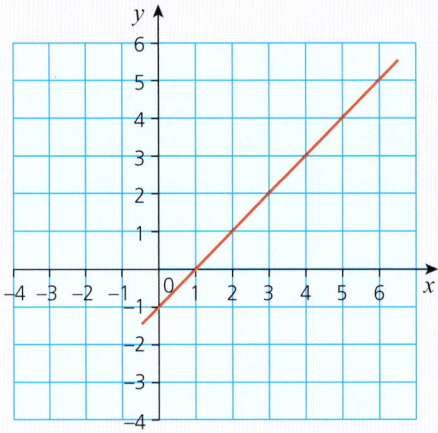

a. Copy and complete the table of coordinates.

x	y
0	
1	
2	
3	
4	

⇒

x	y
0	−1
1	0
2	1
3	2
4	3

b. Describe in words the relationship between the x- and y-coordinates.

The y-coordinate is always 1 less than the x-coordinate.

c. Write your rule using algebra.

$y = x - 1$

d. Does the point with coordinates (10, 9) lie on the line? Justify your answer.

Yes it does. The equation of the line is $y = x - 1$. If we substitute the x- and y-values of the coordinates into the equation, it is correct.

That is $9 = 10 - 1$ ✓

> **KEY INFORMATION**
> If the coordinates did not fit the equation, then the point would not fall on the line.

27 Linear functions

LET'S TALK
Do the values of x and y have to be whole numbers or can they be fractions or decimal values too?

2 An equation of a straight line is given as $y=3x$.

a Copy and complete the table below for the coordinates of some of the points on the line.

x	y		x	y
0			0	0
1			1	3
2			2	6
3			3	9
4			4	12

b Using the table of results, plot the graph of $y=3x$.

The table gives the x- and y-coordinates of five points on the line. These can be plotted and a straight line drawn through them.

KEY INFORMATION
The graph acts as a good check for your table of results above. If one of the points plotted did not lie on the line, then you will have made a mistake in your table of values.

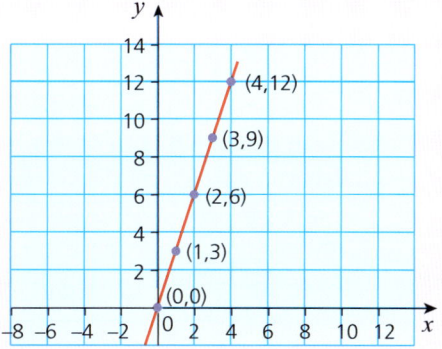

3 The exchange rate between pounds Sterling (£) and US Dollars ($) is given as £1=$1.25.

a Using the number of dollars as D and the number of pounds as P, write a formula linking P and D.

$$D = 1.25P$$

A common error is to write the equation the wrong way around. But as £1=$1.25, you would get more dollars than pounds, so the amount of pounds must be multiplied by 1.25 to work out the amount of dollars.

b Complete the table of results below to convert Pounds to Dollars.

P	D		P	D
0			0	0
1			1	1.25
5			5	6.25
10			10	12.50
20			20	25

SECTION 3

KEY INFORMATION

Note how the result for 10 was given as 12.50 not 12.5.

With currency, decimal values should always be given to two decimal places.

c Plot a graph of D against P for the table of values in part (b).

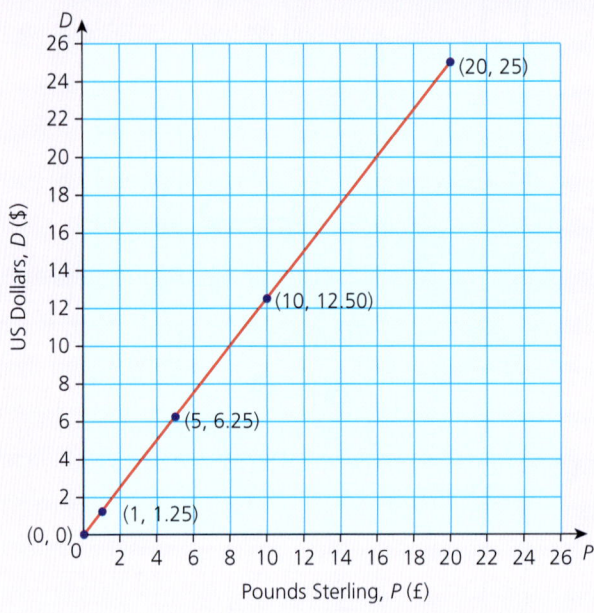

d Use your graph to find an approximate value for the amount of US Dollars you would receive if you exchanged £17.00.

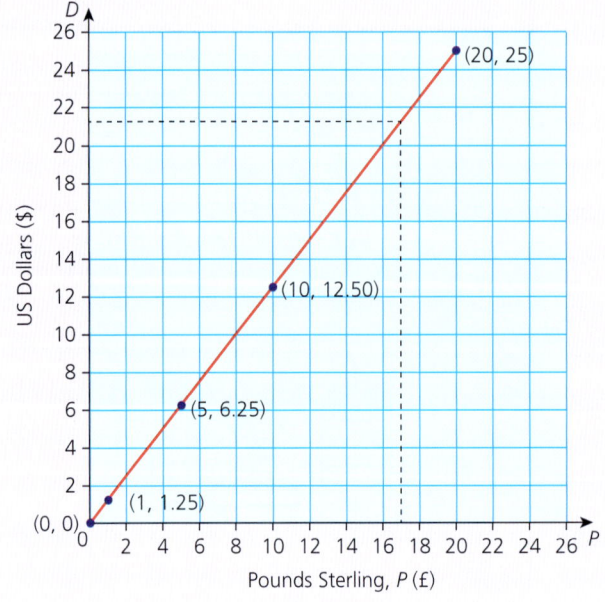

Values taken from a graph are usually only approximate. In this case the actual answer is $21.25, but this cannot be worked out exactly from the graph.

KEY INFORMATION

The symbol ≈ is different from the = sign.

≈ means 'approximately equals to'

From the graph £17 ≈ $21.20

27 Linear functions

Exercise 27.1

1. Look at the graphs below.

   i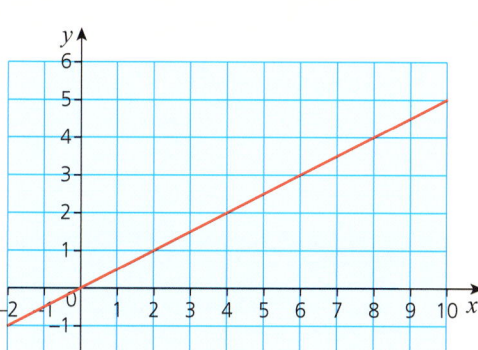

   ii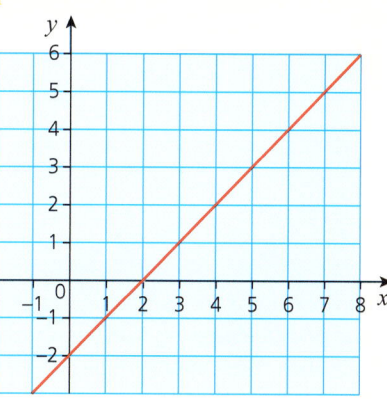

   a For each graph copy and complete the x- and y-coordinate table below.

x	y
0	
2	
4	
6	
8	

   b Describe in words the relationship between the x- and y-coordinates for each of the two graphs.
   c Write equations in the form '$y=$' for each of the two graphs.
   d Which line would the coordinates (50, 48) lie on if the lines were extended? Justify your answer.

2. A line has the equation $y = x + 5$.
   a Which of the coordinates A, B, C and D, given below, lie on the line?
      A(1, 6)   B(–2, 3)   C(0, 5)   D(–4, 1)
   b Give the coordinates of another point that lies on the line.

3. A photocopier manages to print on average 2.5 sheets of paper per second. If t is the time in seconds and n the number of sheets printed:
   a write an equation linking n and t
   b plot a graph of n against t for t ranging from 0 to 10 seconds.

SECTION 3

4 a Copy and complete the coordinate tables below for the two equations $y = x+4$ and $y = \frac{3}{2}x$.

$y = x+4$	
x	y
0	
2	
4	
6	
8	
10	

$y = \frac{3}{2}x$	
x	y
0	
2	
4	
6	
8	
10	

$\frac{3}{2}x$ means that x is being multiplied by $\frac{3}{2}$. $\frac{3}{2}$ as a decimal is the same as 1.5.

b On the same axes plot both graphs.
c One point belongs to both graphs. Explain how you can tell from the coordinate tables, what the coordinates of that point are.

5 The coordinates of seven points are given below.
Three of the points lie only on one line, three of the remaining points lie only on a second line.
One point lies on **both** lines.

P (4, 12)   Q (2, 10)   R (3, 15)   S (8, 40)
T (1, 9)   U (10, 18) V (1, 5)

a Identify which points lie on each of the two lines:
  Line 1 points: ……………………………..…Line 2 points: …………………….…………………
b Which point lies on both lines?
c What are the equations of the two lines?

6 Two taxi firms operate different pricing policies on the amount they charge ($y) for the distance travelled (x km).
Taxi firm A uses the equation $y = 2x$.
Taxi firm B uses the equation $y = x+15$.
The graph below shows the cost of travelling with each of the two taxi firms.

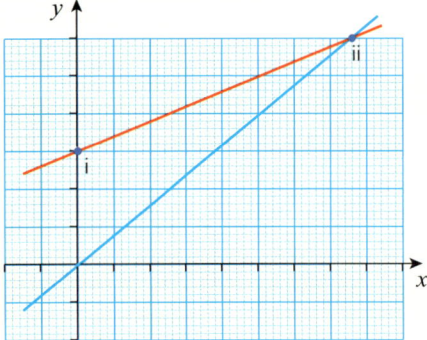

a Copy the graphs and label which one represents taxi firm A and which represents taxi firm B.
b Give the coordinates of the points labelled (i) and (ii). Justify your answers.
c A person wishes to travel in a taxi a total distance of 20 km. Which taxi firm would be the cheapest option? Explain how this can be deduced from the graph above.

27 Linear functions

Horizontal and vertical lines

So far we have only dealt with straight lines which, when plotted, lie diagonally on the axes. The equations of these lines are written with two variables, e.g. x and y.

But straight lines can be drawn horizontally or vertically too.

Worked examples

1. By looking at the coordinates of some of the points on the line below, establish the equation of the straight line.

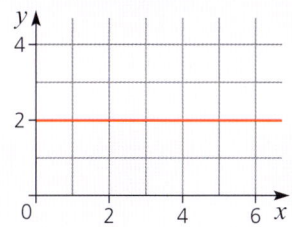

x	y
0	2
2	2
4	2
6	2

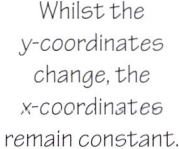

Whilst the x-coordinates change, the y-coordinates remain **constant**.

By looking at the table it can be seen that the only rule that all the points have in common is that the y-values are always equal to 2. Therefore the equation of the straight line is $y = 2$.

2. By looking at the coordinates of some of the points on the line below, establish the equation of the straight line.

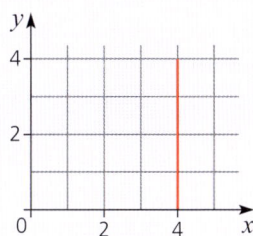

x	y
4	0
4	1
4	2
4	3

Whilst the y-coordinates change, the x-coordinates remain constant.

By looking at the table it can be seen that the only rule that all the points have in common is that the x values are always equal to 4. Therefore the equation of the straight line is $x = 4$.

217

SECTION 3

Exercise 27.2

1. In each of the following identify the coordinates of some of the points on the line and use these to find the equation of the straight line.

a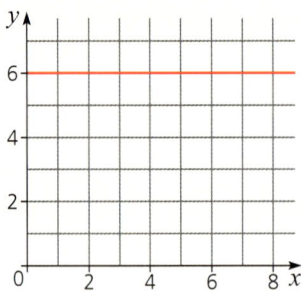

e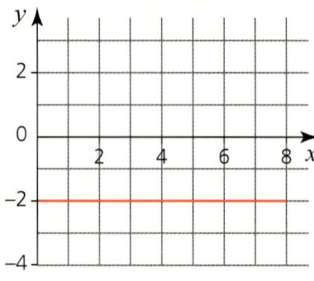

b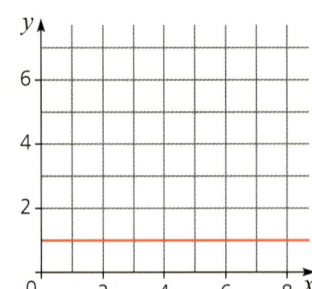

f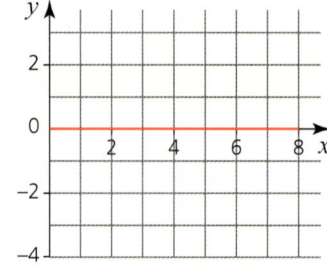

c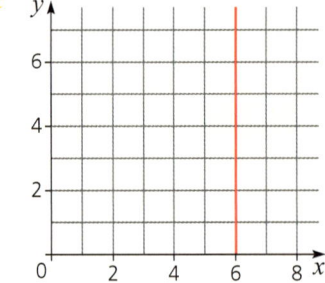

g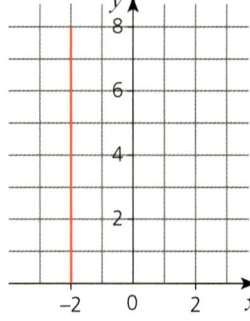

d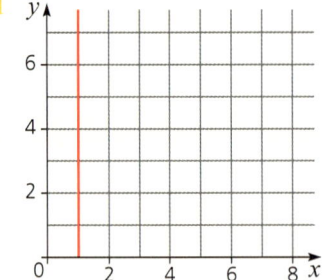

h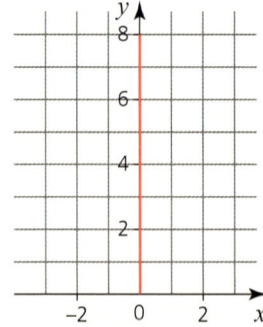

27 Linear functions

2 In the following questions, work out, just by looking at the equation, whether if plotted it would produce either a horizontal or vertical line.
- a $y=6$
- b $x=3$
- c $x=-3$
- d $y=-2$
- e $x=6$
- f $y=-3$
- g $x=0$

LET'S TALK
Is it possible to tell which lines are parallel to each other simply by their equation? How do you know?

3 The equations of two straight lines l_1 and l_2 are given as:
l_1: $y=6$
l_2: $x=8$
The coordinates of six points are given below:
A(2, 6)   B(−4, 6)   C(8, 6)   D(0, 6)   E(8, 0)   F(8, −6)
- a Work out which points lie on which line.
- b Is there a point which lies on both lines? Justify your answer.

4 A rectangle ABCD is shown on the axes below.

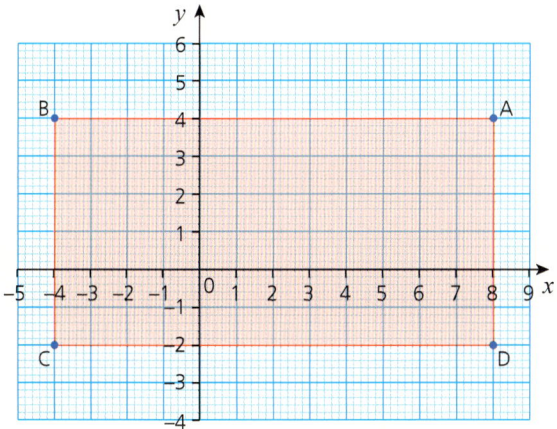

Calculate the equation of the lines passing through the following vertices:
- a A and B
- b B and C
- c A and C

5 An art gallery has a small plot of land with 10 sculptures.
The sculptures are shown as points on the grid below.
The distances are measured in metres.

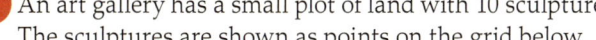

SECTION 3

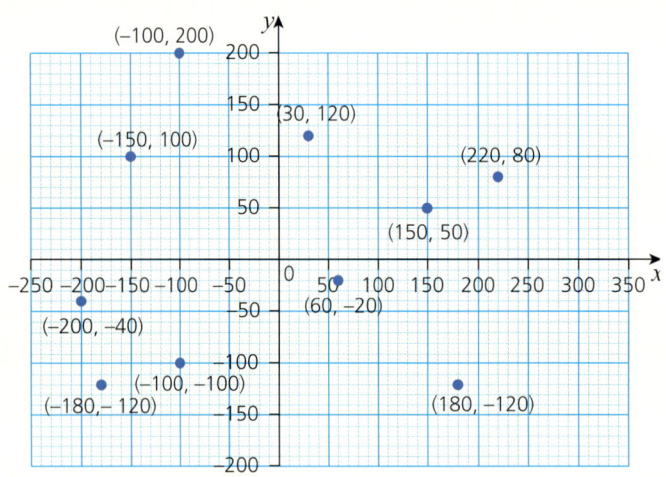

a The gallery is building a rectangular enclosure to keep the sculptures safe.
   It is planning to build straight fences along the lines $y=160$, $y=-140$, $x=210$ and $x=-190$ to form the rectangular enclosure.
   Would any sculpture(s) be outside the enclosure? Justify your answer.
b Give the equations of the fence lines which would form the smallest possible rectangular enclosure that has all 10 sculptures inside. (Assume a sculpture can lie on a fence line.)
c What is the area of the smallest possible enclosure that the gallery can build?

6 The grid below shows vertical lines V_1, V_2, V_3 and V_4 with equations $x=8$, $x=16$, $x=24$ and $x=32$ respectively.
   It also shows horizontal lines H_1, H_2, H_3 and H_4 with equations $y=6$, $y=12$, $y=18$ and $y=24$ respectively.

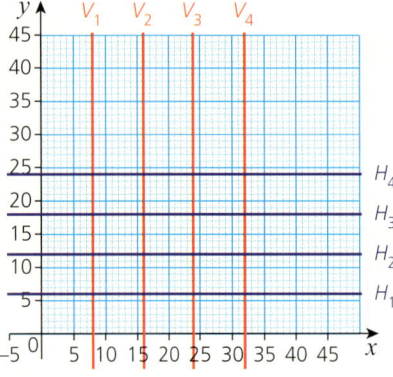

a Assuming the sequence of horizontal and vertical lines continues as shown, deduce the equation of lines H_{10} and V_{15}.
b What are the coordinates of the **point of intersection** of the lines H_{20} and V_{30}?

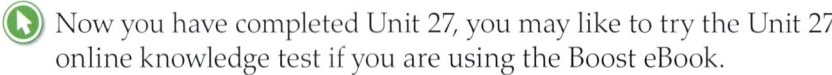

Now you have completed Unit 27, you may like to try the Unit 27 online knowledge test if you are using the Boost eBook.

28 Converting units and scale drawings

- Understand the relationships and convert between metric units of area, including hectares (ha), square metres (m²), square centimetres (cm²) and square millimetres (mm²).
- Use knowledge of scaling to interpret maps and plans.

Some history

In one day a soldier in Julius Caesar's army could comfortably march 20 miles wearing full kit, and then help to build a defensive stockade.

The mile was a unit of length based upon 1000 strides of a Roman legionary.

The measurement was sufficiently accurate for its purpose but only an approximate distance.

Most measures started as rough estimates. The yard (3 feet or 36 inches) was said to be the distance from the English king's nose (reputed to be Edward I) to the tip of his extended finger.

> **LET'S TALK**
> How could you estimate how far 1000 of your strides is?

> **KEY INFORMATION**
> A standardised measurement is a fixed quantity which is exact.

As it became necessary to have standardisation in measurement, the measures themselves became more exact.

The metric system

The metric system uses a number of units for length. They are:

kilometre (km), metre (m), centimetre (cm) and millimetre (mm)

SECTION 3

KEY INFORMATION
centi comes from the Latin 'centum' meaning hundred, milli comes for the Latin 'mille' meaning thousand, kilo comes from the Greek 'khilioi' meaning thousand.

The units for mass are:

tonne (t), kilogram (kg), gram (g) and milligram (mg)

The units for capacity are:

litre (l) and millilitre (ml)

Exercise 28.1

1. Copy and complete the sentences below.
   a. There are _____ centimetres in one metre.
   b. A centimetre is _____ part of a metre.
   c. There are _____ metres in one kilometre.
   d. A metre is _____ part of a kilometre.
   e. There are _____ grams in one kilogram.
   f. A gram is _____ part of a kilogram.
   g. A kilogram is _____ part of a tonne.
   h. There are _____ millilitres in one litre.
   i. One thousandth of a litre is _____.
   j. There are _____ grams in one tonne.

2. Which of the units below would you use to measure each of the following?
   mm   cm   m   km   mg   g   kg   tonne   ml   litre
   a. your mass
   b. the length of your foot
   c. your height
   d. the amount of water in a glass
   e. the mass of a ship
   f. the height of a bus
   g. the capacity of a swimming pool
   h. the length of a road
   i. the capacity of the fuel tank of a truck
   j. the size of your waist

3. Write an estimate for each of the following using a sensible unit.
   a. your height
   b. your mass
   c. the capacity of a cup
   d. the distance to the nearest town
   e. the mass of an orange
   f. the quantity of blood in the human body
   g. the depth of the Pacific Ocean
   h. the distance to the Moon
   i. the mass of a car
   j. the capacity of a swimming pool

LET'S TALK
Carry out some research to check your estimates. Have all the quantities got one definitive value? Explain your answer.

Converting from one unit to another

Length

1 km is 1000 m, so
- to change from km to m, multiply by 1000
- to change from m to km, divide by 1000.

> **Worked examples**
>
> 1 Change 5.84 km to metres.
>   1 km = 1000 m, so multiply by 1000.
>   5.84 × 1000 = 5840 m
> 2 Change 3640 mm to metres.
>   1000 mm = 1 m, so divide by 1000.
>   3640 ÷ 1000 = 3.64 m

Mass

1 tonne is 1000 kg, so
- to change from tonnes to kg, multiply by 1000
- to change from kg to tonnes, divide by 1000.

> **Worked examples**
>
> 1 Change 0.872 tonne to kilograms.
>   1 tonne = 1000 kg, so multiply by 1000.
>   0.872 × 1000 = 872 kg
> 2 Change 4200 kg to tonnes.
>   1000 kg = 1 tonne, so divide by 1000.
>   4200 ÷ 1000 = 4.2 tonnes

Capacity

1 litre is 1000 ml, so
- to change from litres to ml, multiply by 1000
- to change from ml to litres, divide by 1000.

SECTION 3

> **Worked examples**
>
> 1. Change 2.4 litres to millilitres.
>   1 litre is 1000 ml, so multiply by 1000.
>   2.4 × 1000 = 2400 ml
> 2. Change 4500 ml to litres.
>   1000 ml is 1 litre, so divide by 1000.
>   4500 ÷ 1000 = 4.5 litres

Exercise 28.2

1. Copy and complete the sentences below.
   a. 1 m is _____ cm, so
      to change from m to cm _____
      to change from cm to m _____.
   b. 1 m = _____ mm, so
      to change from m to mm _____
      to change from mm to m _____.
   c. 1 cm = _____ mm, so
      to change from cm to mm _____
      to change from mm to cm _____.

2. Convert these to millimetres.
   a. 4 cm
   b. 6.2 cm
   c. 28 cm
   d. 1.2 m
   e. 0.88 m
   f. 3.65 m
   g. 0.008 m
   h. 0.23 cm

3. Convert these to metres.
   a. 260 cm
   b. 8900 cm
   c. 2.3 km
   d. 0.75 km
   e. 250 cm
   f. 0.4 km
   g. 3.8 km
   h. 25 km

4. Convert these to kilometres.
   a. 2000 m
   b. 26 500 m
   c. 200 m
   d. 750 m
   e. 100 m
   f. 5000 cm
   g. 15 000 cm
   h. 75 600 mm

5. Copy and complete the sentences below.
   1 kg is _____ g, so
   to change kg to g _____
   and to change g to kg _____.

6. Convert these to kilograms.
   a. 2 tonnes
   b. 7.2 tonnes
   c. 2800 g
   d. 750 g
   e. 0.45 tonnes
   f. 0.003 tonnes
   g. 6500 g
   h. 7 000 000 g

7. Convert these to millilitres.
   a. 2.6 litres
   b. 0.7 litre
   c. 0.04 litre
   d. 0.008 litre

28 Converting units and scale drawings

8 Convert these to litres.
    a 1500 ml   **b** 5280 ml   **c** 750 ml   **d** 25 ml

9 How much water is left in a 1-litre bottle after 400 ml is drunk?

10 How much water is left in a 1.5-litre bottle after 750 ml is drunk?

11 A pile of bricks has a mass of 16 kg.
9450 g of bricks are removed. What mass is left?

12 I am travelling to a city 250 km away.
I take a break after 173 000 m. How many kilometres do I have left to travel?

13 A cyclist travels for five days and covers 375 000 m in total.
He covers 67 km, 78 km, 46 km and 89 km on the first four days.
How far does he travel on day 5?

14 The masses of four containers loaded on a ship are 28 tonnes, 45 tonnes, 16.8 tonnes and 48 500 kg. What is the total mass in tonnes?

15 Three test tubes contain 0.08 litre, 0.42 litre and 220 ml of solution.
    a What is the total volume of solution in millilitres?
    b How many litres of water need to be added to make the volume of solution up to 1.25 litres?

16 A company sends the same parcel to each of its customers.
On average the company sends out 150 parcels a week.
Each parcel uses 40 cm of string.
A ball of string costs $2.25 and has string wrapped around it of total length 50 m.
Estimate how much the company pays per year to buy the string for wrapping. Show all your working clearly.

Units of area

We can convert between different units of area.

1 cm = 10 mm, so

$$1\,cm^2 = 10\,mm \times 10\,mm = 100\,mm^2$$

Similarly, 1 m = 100 cm, so

$$1\,m^2 = 100\,cm \times 100\,cm = 10\,000\,cm^2$$

and 1 m = 1000 mm, so

$$1\,m^2 = 1000\,mm \times 1000\,mm = 1\,000\,000\,mm^2$$

Also 1 km = 1000 m, therefore

$$1\,km^2 = 1000\,m \times 1000\,m = 1\,000\,000\,m^2$$

SECTION 3

> The abbreviation for hectare is ha.

Another unit of area is the **hectare**.

A hectare is an area equivalent to the area of a square of side length 100 m.

Therefore 1 ha = 100 m × 100 m = 10 000 m²

Exercise 28.3

1. Copy and complete the following.
   a. 5 cm² = _____ mm²
   b. 8 m² = _____ cm²
   c. _____ m² = 7.2 ha
   d. 64 000 cm² = _____ m²
   e. _____ cm² = 5600 mm²

2. Write an estimate for each of the following using a sensible unit.
   a. The area of a football field
   b. The area of a stamp
   c. The area of a city

LET'S TALK
Carry out some research to check your estimates.

3. A bathroom is in the shape of a cuboid with dimensions as shown.

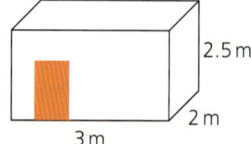

The door is 80 cm × 200 cm.
Square wall tiles are 10 cm × 10 cm and come in boxes of 50.
Each box costs $12.50.
Assuming only whole boxes can be bought. Calculate the cost of tiling the four walls of the bathroom. Show all your working clearly.

Scale drawings

Sometimes drawing an object to its actual size is not practical or possible. When architects design buildings, for example, they do not produce a life-size drawing of the building. A scale drawing is produced instead. In a scale drawing, all the lengths are changed by the same scale factor, covered in Unit 13.

The scale is shown using a **ratio**. For example, a scale of 1 : 50 means that 1 unit of length on the diagram represents a length which is 50 times bigger on the actual object. Therefore 1 cm on the diagram represents 50 cm in real life. Ratios will be explored fully in Unit 29.

28 Converting units and scale drawings

> **Worked examples**
>
> 1. A diagram is drawn to a scale of 1:25. A line on the diagram has length 20 cm. What actual length does it represent?
>
>   1:25 means that the actual lengths are 25 times bigger than those drawn.
>
> $20 \times 25 = 500$
>
>   So 20 cm on the diagram represents 500 cm (5 m) in real life.
>
> 2. This sketch shows the dimensions of a room. Draw a scale diagram of the room with a scale of 1:150.
>
>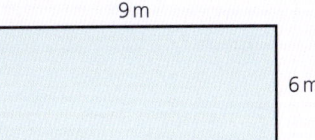
>
>   1:150 means that the lengths on the diagram must be 150 times smaller than those of the actual room. We need to divide by 150.
>
>   9 m = 900 cm so the diagram length is 900 cm ÷ 150 = 6 cm.
>
>   6 m = 600 cm so the diagram length is 600 cm ÷ 150 = 4 cm.
>
>   The scale drawing is as shown.
>
>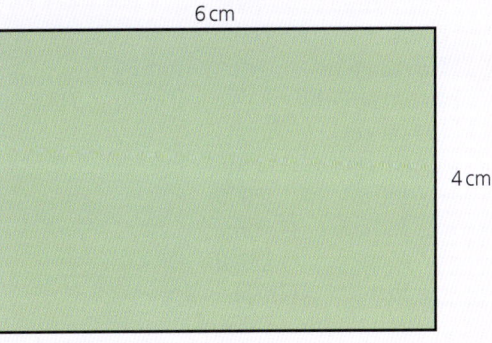
>
> 3. A circular arena is being built. On a scale drawing, the radius of the arena is 8 cm. The radius of the actual arena is 120 m. Calculate the scale of the drawing.
>
>   To calculate the scale factor of enlargement from the diagram to the actual arena, both lengths must be in the same units.
>
>   On the diagram, radius = 8 cm.
>
>   In real life, radius = 120 m = 12 000 cm.
>
>   Scale factor of enlargement = 12 000 ÷ 8 = 1500.
>
>   Therefore the scale of the drawing is 1:1500.

Parts of a circle were taught in Unit 2.

SECTION 3

4   A landscape artist makes a scale drawing of a public garden. The scale used is 1 cm to represent 10 m.

   a   The width of the garden on the scale drawing is 18.2 cm. Calculate its real width.

   1 cm represents 10 m, so 18.2 cm represents $18.2 \times 10 = 182$ m.

   The real width is 182 m.

   b   The length of the actual garden is 425 m. Calculate its length on the scale drawing.

   10 m is represented by 1 cm, so 425 m is represented by $425 \div 10 = 42.5$ cm.

   The length on the drawing is 42.5 cm.

Exercise 28.4

1   Calculate the actual length (in metres) represented by each of these lengths on a scale drawing. The scale of each diagram is given in brackets.
   a   10 cm   (1:50)
   b   35 cm   (1:30)
   c   40 mm   (1:100)
   d   160 mm   (1:150)
   e   70 cm   (2:75)

2   Calculate the length (in centimetres) that represents each of these actual lengths on a scale drawing. The scale of each diagram is given in brackets.
   a   60 m   (1:150)
   b   120 m   (1:2000)
   c   380 cm   (1:25)
   d   4 km   (1:5000)
   e   15 m   (2:75)

3   In each of these pairs of lengths, the first is the length on a scale drawing and the second is the corresponding length in real life. Calculate the scale in each case.
   a   20 cm   1200 cm
   b   25 cm   100 m
   c   5 cm   75 m
   d   6 cm   0.3 km
   e   12 mm   600 m

4   Each of these sketches shows the plan view of a room and its actual dimensions.
   i)   Draw a scale diagram of each room using the scale suggested.
   ii) Write the dimensions (in centimetres) on each diagram.

28 Converting units and scale drawings

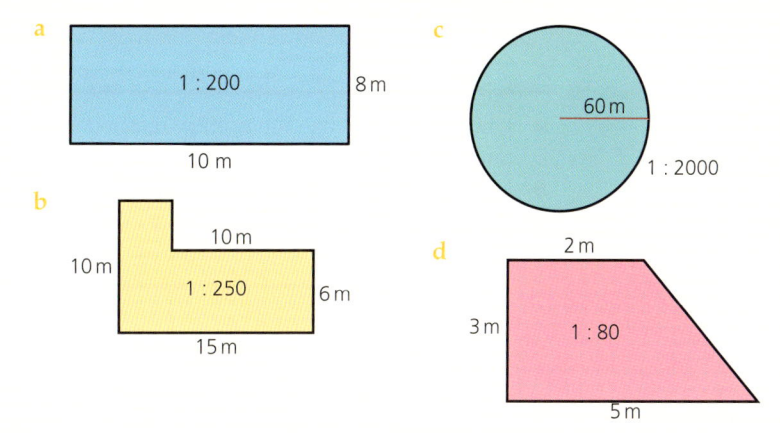

5. Draw a scale diagram of your classroom.
   You will need a tape measure, pencil and squared paper.

6. The ground floor of a building is drawn to scale. 1 cm on the drawing represents 10 m in real life.
   a The length is 8.3 cm on the scale drawing. Calculate the length of the real building.
   b The width of the actual building is 128 m. Calculate its width on the drawing.

7. This diagram shows the plan of a room drawn to a scale of 1:50.

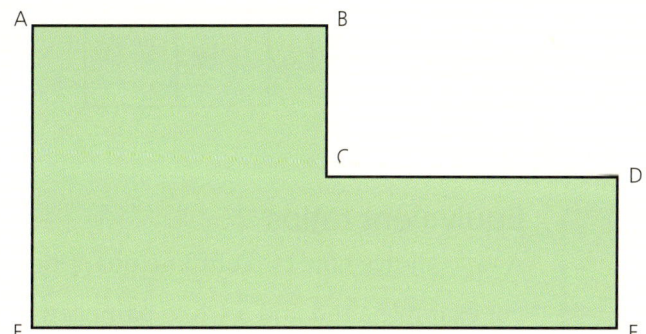

   Calculate the area of the room in m².

8. Using an atlas or the internet, find a map of your local area.
   a What is the scale of the map that the map uses?
   b Choose two towns or cities on your map and estimate the actual distance between them in kilometres.
   c If you were to cycle that distance, estimate how long it would take you.
   You must show all your workings clearly.

KEY INFORMATION

Work in pairs on question 8.

For part (c) you will need to explain clearly what assumptions you have made to work out the answer.

Now you have completed Unit 28, you may like to try the Unit 28 online knowledge test if you are using the Boost eBook.

29 Ratio

- Understand and use the unitary method to solve problems involving ratio and direct proportion in a range of contexts.
- Use knowledge of equivalence to simplify and compare ratios (same units).
- Understand how ratios are used to compare quantities to divide an amount into a given ratio with two parts.

Fractions and ratios

Equivalent fractions

We can see from the diagram below that $\frac{1}{2}$, $\frac{2}{4}$ and $\frac{4}{8}$ are **equivalent fractions**.

They are called this because each fraction is worth the same amount.

Similarly $\frac{1}{3}$, $\frac{2}{6}$ and $\frac{3}{9}$ are equivalent fractions, as are $\frac{1}{5}$, $\frac{10}{50}$ and $\frac{20}{100}$.

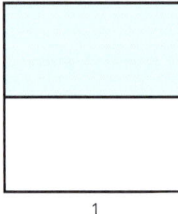
$\frac{1}{2}$

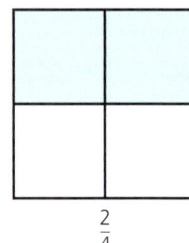
$\frac{2}{4}$

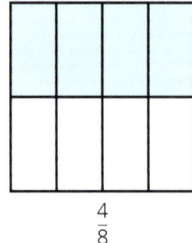
$\frac{4}{8}$

> **LET'S TALK**
> If the pink paint was mixed in the ratio 1 : 5, would the paint be darker or lighter than the one mixed in the ratio 1 : 4?

Equivalent ratios

A ratio shows how two or more quantities are related to each other.

For example a red and a white paint can be mixed in a ratio of 1 : 4 to make a pink paint.

This means that for every 1 unit of red paint, there are 4 units of white paint, i.e. there is four times as much white paint than red in the mixture that makes pink.

Ratios behave in a similar way to fractions.

1 : 2 is equivalent to 2 : 4 or 35 : 70.

In the same way, 15 : 5 is equivalent to 3 : 1 or 9 : 3.

29 Ratio

Exercise 29.1

Copy these sets of equivalent ratios and fill in the blanks.

1) $4:5 = 8:\square = \square:50 = 12:\square$

2) $7:2 = 14:\square = \square:10 = 49:\square$

3) $8:5 = \square:50 = 32:\square = 4:\square$

Ratios, like fractions, can also be simplified. To write a ratio in its **simplest form**, simply divide by the highest common factor of the numbers involved.

Worked example

A tile pattern is made of two different coloured tiles, red and blue.

The ratio of red:blue tiles is $50:75$. Write this ratio in its simplest form.

The highest common factor of both 50 and 75 is 25, i.e. 25 is the largest number that goes in to both 50 and 75.

$$50:75$$
$$\div 25 \quad \quad \div 25$$
$$2:3$$

The ratio of red:blue tiles in its simplest form is $2:3$.

KEY INFORMATION
When writing a ratio in its simplest form, the numbers must be whole numbers, not decimals or fractions. In this example, $2:3$ cannot be simplified further to $1:1.5$.

Some ratios, however, are written in the form $1:n$, where n is a whole number or a decimal.

For example, in a school the teacher-to-student ratio may be given as $1:15.5$. This means that for every teacher, there are 15.5 students. (It does not mean that every class has 15.5 students but that, if the number of students is divided by the number of teachers, the answer is 15.5.)

SECTION 3

Worked examples

1. A school has 40 teachers and 720 students.

   Write the teacher:student ratio in the form $1:n$.

   The teacher:student ratio is $40:720$.

   Therefore to write $40:720$ in the form $1:n$ and keep the ratios equivalent, both sides must be divided by the same number.

   Teachers : Students

   Therefore $n = 720 \div 40$
   $= 18$

   The ratio of teachers:students in the form $1:n$ is $1:18$.

2. The teacher:student ratio in a school is $1:18$.

   There are 25 teachers. How many students are there?

   The ratio is given in the form $1:n$ so the values need to be multiplied to keep the ratio constant.

   Teachers : Students

   $1 : 18$
   $\times 25 \quad \quad \times 25$
   $25 : 450$

   Therefore the number of students $= 18 \times 25$
   $= 450$

Exercise 29.2

1. A school has 20 teachers and 480 students.
   What is the teacher:student ratio in its simplest form?
2. A college has 250 staff and 3500 students.
   What is the staff:student ratio in its simplest form?
3. A town in America has 2400 families and 4200 cars.
   What is the family:car ratio
   a. in its simplest form
   b. in the form $1:n$.
4. An alloy contains copper and tin in the ratio $3:1$.
   40 g of tin is used to make the alloy.
   How much copper is used?

5. The pie chart below shows the proportion, in degrees, of the different colour hair of pupils in a class.

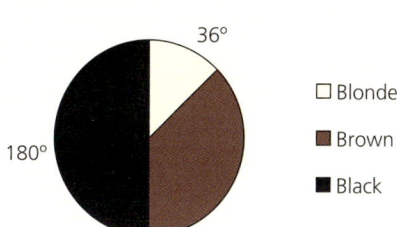

Write the ratio of Blonde : Brown : Black in its simplest form.

KEY INFORMATION
Note: one dozen is 12.

6. In a batch of 10 dozen eggs, 25 are cracked.
   a What is the ratio of cracked eggs to un-cracked eggs in:
      i) its simplest form
      ii) in the form $1:n$.
   b If this ratio is the average for all batches, how many cracked eggs would you expect to find in four dozen eggs?
   c A large catering company needs 360 un-cracked eggs for 1 week of catering. How many eggs will they need to buy to ensure they have enough?

7. A tile pattern is made up of three different coloured tiles, red, blue and yellow as shown.

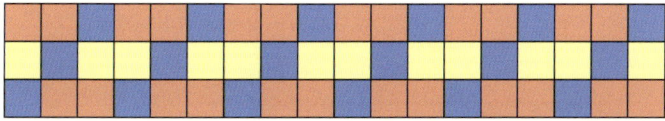

   a Write the ratio of yellow : blue : red tiles in its simplest form.
   b If a complete tile pattern has 320 yellow tiles, how many blue and red tiles will it have?
   c If a complete tile pattern has 423 blue tiles, how many yellow and red tiles will it have?

8. Luisa and Pedro are sister and brother. They decide to save their pocket money in their money boxes as follows:
   Luisa starts by putting in $15 then adds $3 to it each week.
   Pedro starts by putting in only $5, but then adds $4 to it each week.
   After how many weeks will the ratio of their savings be $1:1$?

9. P, Q and R represent three whole numbers.
   The ratio of $P:Q$ is $3:5$, whilst the ratio of $Q:R$ is $9:10$.
   What is the smallest possible value of $P+Q+R$?

SECTION 3

LET'S TALK

If you plotted a graph of two variables that are in direct proportion to each other (for example, number of hours worked plotted against pay), what would the graph look like? What properties does it have?

Direct proportion

Workers in a shop are paid for the number of hours they work. Their pay is in **direct proportion** to the number of hours they work – more hours, more pay.

Workers producing cups on a machine in a factory are sometimes paid for the number of cups they produce, not for the time they work. Their pay is in direct proportion to the number of cups they make.

Worked example

A machine for making bread rolls makes 1500 rolls in 20 minutes.

How many rolls will it make in 3 hours?

Let x be the number of rolls made in 3 hours.

Minutes : Rolls

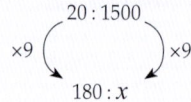

20 : 1500

×9 ⤵ ⤵ ×9

180 : x

The ratios have to be equivalent. You multiply the number of minutes by 9 to get 180 so you must also multiply the number of rolls by 9.

$1500 \times 9 = 13\,500$, so 13 500 rolls are made in 3 hours.

Note:
3 hours = 180 minutes. This conversion from hours to minutes must be done so that all the units of time are the same

Exercise 29.3

LET'S TALK

How do you convert the answer to question 4(a) from a decimal into hours and minutes?

1. A heater uses 3 units of electricity in 40 minutes.
   How many units does it use in 2 hours?
2. A machine prints 1500 newspapers in 45 minutes.
   How many does it print in 12 hours?
3. A bricklayer lays 1200 bricks in an average 8-hour day.
   How many bricks does he lay in a 40-hour week?
4. A machine puts tar on a road at the rate of 4 metres in 5 minutes.
   a How long does it take to cover 1 km of road?
   b How many metres of road does it cover in 8 hours?

5 Gaby uses the following recipe to make 12 chocolate shortbread biscuits.

> 260 g of plain flour
>
> 100 g of sugar
>
> 40 g of cocoa powder
>
> 200 g of unsalted butter

a How much of each quantity does Gaby need to make 15 biscuits?
b If Gaby only has 200 g of plain flour, what is the maximum number of biscuits she can make? (Assume she has enough of each of the other ingredients.)
c Another time Gaby wants to make these biscuits, she checks her ingredients and finds she has the following amounts.
1 kg of plain flour   400 g of sugar
150 g of cocoa powder   650 g of unsalted butter
She wants to make 42 of the biscuits.
  i) Has she got enough ingredients? Justify your answer.
  ii) What is the maximum number of biscuits she can make?

When the information is given as a ratio, the method of solving the problem is the same.

Worked example

Copper and nickel are mixed in the ratio $5:4$.

48 g of nickel is used. How much copper is used?

Let x grams be the mass of copper needed.

Copper : Nickel

$$\begin{array}{c} 5:4 \\ \times 12 \Big(\quad\Big) \times 12 \\ x:48 \end{array}$$

The ratios have to be equivalent. You multiply the mass of nickel by 12 to get 48 g so you must also multiply the mass of copper by 12.

$5 \times 12 = 60$, so 60 g of copper is used.

SECTION 3

Exercise 29.4

1. The ratio of girls to boys in a class is 6:5.
   There are 18 girls. How many boys are there?
2. Sand and gravel are mixed in the ratio 4:3 to make ballast. 80 kg of sand is used. How much gravel is used?
3. A paint mix uses blue and white in the ratio 3:10.
   6.6 litres of blue paint are used. How much white paint is used?
4. A necklace has green and blue beads in the ratio 2:3.
   There are 24 green beads on a necklace. How many blue beads are there?
5. The size of the three angles, x, y and z, of a triangle are in the ratio 1:3:5 respectively.
   Calculate the size of each of the angles.
6. Three colours of paint, white, blue and orange, are mixed in a certain ratio.
   Part of this ratio is shown below as a ratio and part shown as a pie chart in degrees.

Paint mix

> You may wish to see what final colour this ratio of paints makes. How does changing the ratio of the three colours affect the final colour?

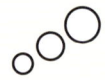

White : Blue : Orange
  z : 5 : 8
Calculate the values of x, y and z.

Now you have completed Unit 29, you may like to try the Unit 29 online knowledge test if you are using the Boost eBook.

30 Graphs and rates of change

- Read and interpret graphs related to rates of change. Explain why they have a specific shape.

Many real-life situations can be represented by graphs. This unit will look at some of those types of graphs and more importantly how to interpret them.

Travel graphs

When an object moves, there are several variables that can be measured. These include the distance it travels and the speed it is travelling at. The object's movement can be represented and studied using a distance–time graph.

Worked example

The motion of a radio-controlled car travelling in a straight line is shown on the graph below.

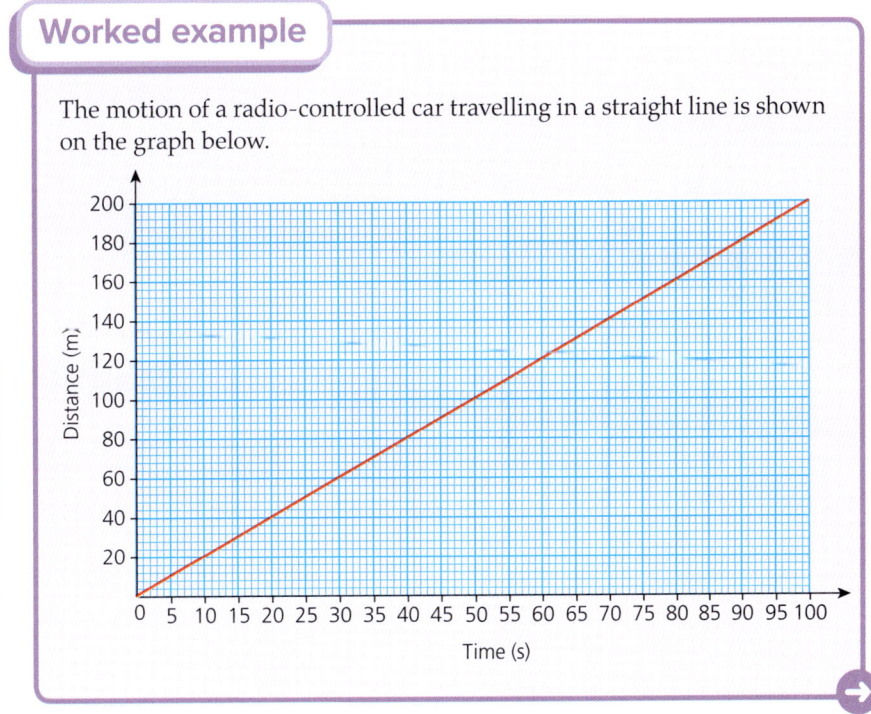

> A constant rate of change of distance with time means that the speed is constant.

SECTION 3

LET'S TALK
If the car travelled at a faster constant speed, how would the graph change? Justify your answer.

a Describe the motion of the car and justify your answer.

The distance travelled is increasing at a constant rate.

This means that the car's speed is constant.

This is shown by the graph being a straight line.

b How far had the car travelled after 1 minute?

From the graph a line is drawn up from 60 seconds until it meets the graph and then drawn across to meet the y-axis as shown.

After 1 minute the car has travelled 120 metres.

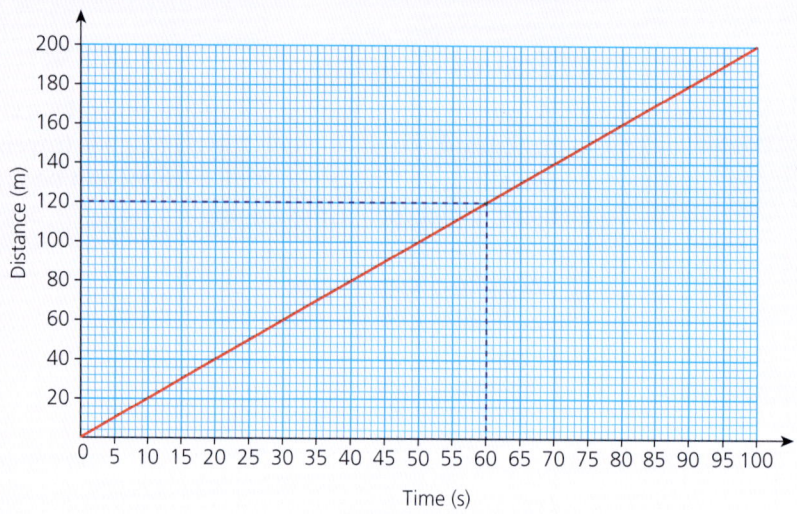

If the speed is not constant, then an average speed can be calculated instead by dividing the total distance travelled by the total time taken.

KEY INFORMATION
As the unit of distance is metres and the unit of time is seconds, the units of speed are metres per second. This is written as m/s.

c How fast is the car travelling?

The speed of an object can be calculated by dividing the distance travelled by the time it took to travel that distance, so $speed = \frac{distance}{time}$.

In this case, because the car is travelling at a constant speed, any part of the graph can be used to calculate the speed.

Using the values worked out in part (b) above:
$speed = \frac{120}{60} = 2 \, m/s$.

30 Graphs and rates of change

Exercise 30.1

1. The three distance–time graphs below show the motion of a car over time.

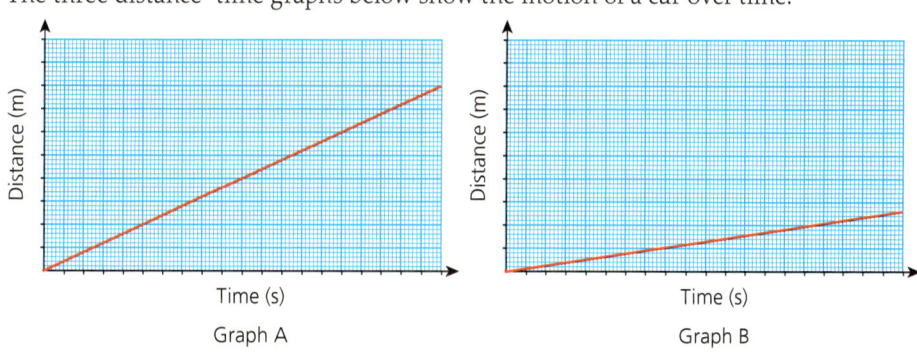

Graph A

Graph B

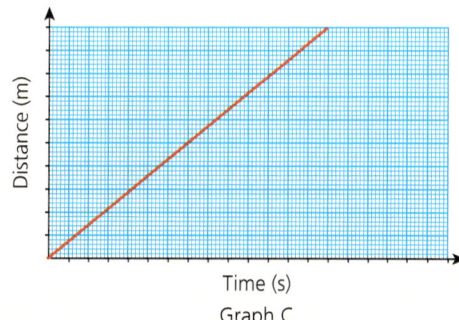

Graph C

Assuming the three graphs are drawn to the same scale, which one shows the car travelling the slowest? Justify your answer.

2. The distance–time graph below shows the motion of a train over a period of time.

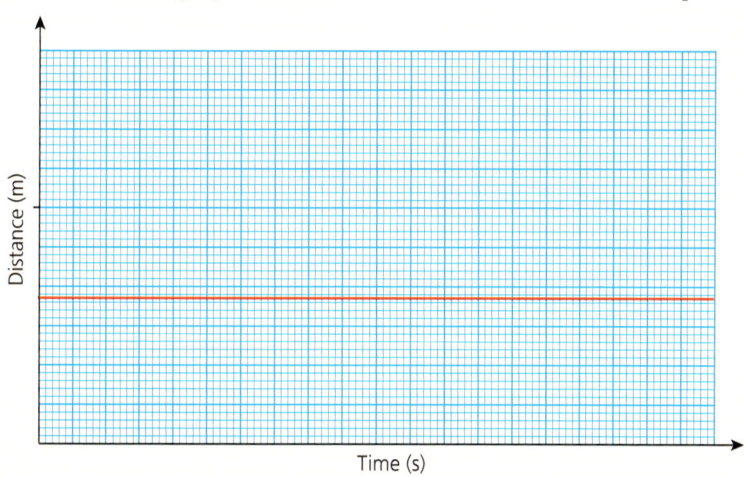

SECTION 3

Choose the statement below which best describes the motion of the train. Justify your choice.

i) The train is travelling at a constant speed.
ii) The train has stopped.
iii) The train is travelling on flat ground.
iv) The train is travelling in a straight line.

LET'S TALK
Discuss why the graph does not necessarily show the three statements you rejected.

3 The distance–time graph below shows the motion of a cyclist over a period of time.

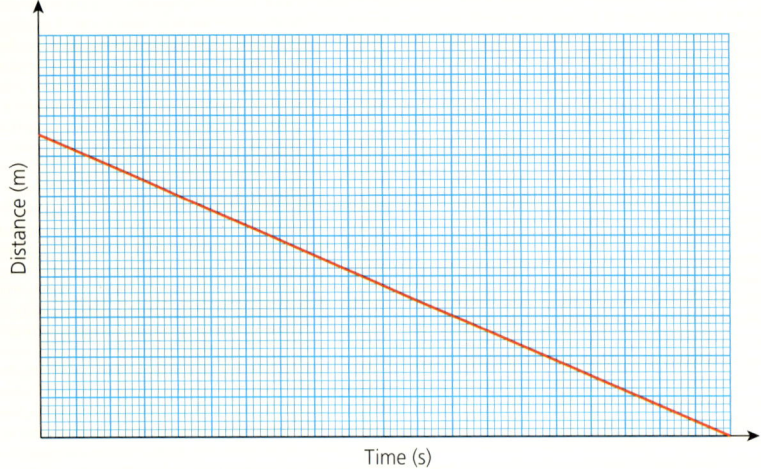

Choose the statement(s) below which describes the motion of the cyclist. Justify your choice(s).

i) The cyclist is travelling at a constant speed.
ii) The cyclist is travelling back to a point.
iii) The cyclist is riding down a hill.
iv) The cyclist is slowing down.

LET'S TALK
Discuss why the graph does not necessarily show the statements you rejected.

4

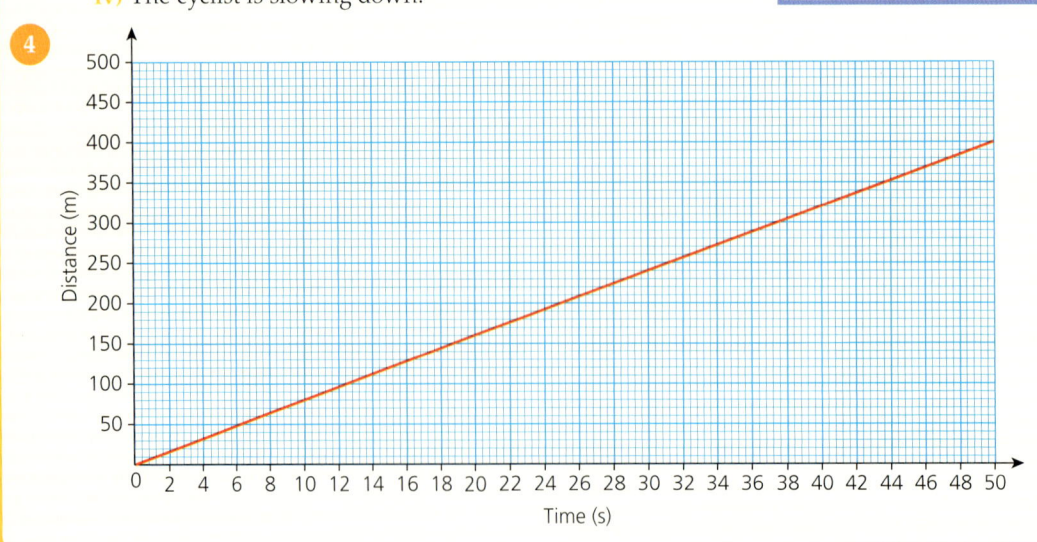

The graph above shows an object travelling at 8 m/s (a distance of 8 m each second).
Using the graph or by any other method, find:
  a how long the object takes to travel 50 m
  b how far the object travels in 5 seconds
  c how long the object takes to travel 70 m
  d how far the object travels in 8 seconds
  e how long the object takes to travel 650 m
  f how far the object travels in 1 minute
  g how long the object takes to travel 1 km
  h how far the object travels in 5 minutes
  i how long the object takes to travel 550 m
  j how many metres the object travels in 1 hour.

5 An electric car travels at 12 m/s in a straight line.
   Draw a graph of the first 10 seconds of its motion.

6 The graph below shows a family car journey.

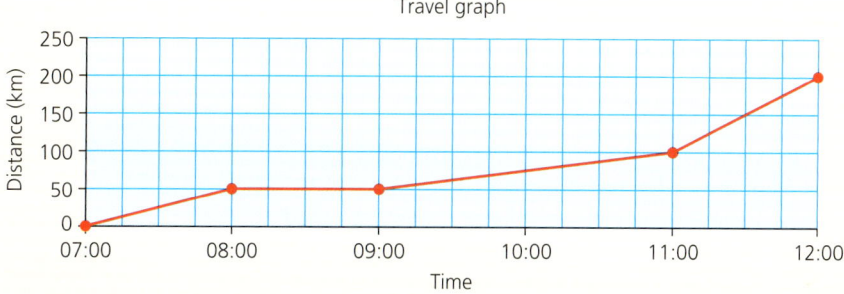

  a What time did the family set out?
  b i) How far did they travel in the first hour?
     ii) What was their average speed?
  c What time did they stop for breakfast?
  d How long did they stop for breakfast?
  e How far did they travel between 09:00 and 11:00?
  f Calculate their average speed during those two hours.
  g The family reached a motorway at 11.00. What was their average speed for the next hour?
  h How long did the whole journey take?
  i How far did they travel in total?
  j What was the average speed for the whole journey?

7 A salesman leaves home at 08:00 and travels for $1\frac{1}{2}$ hour at an average speed of 60 km/h. He
   then stops for 30 minutes. He continues for 2 more hours at 50 km/h and stops for 1 hour.
   He then returns home and arrives home at 16:00.
   a Show this in a travel graph.
   b Calculate his average speed for his return journey.

SECTION 3

8 A train leaves a station at 07:30. It travels for 1 hour 30 minutes at 100 km/h and then stops for 15 minutes. It travels a further 200 km at 100 km/h and then stops for 30 minutes. It then does the return journey non-stop at 100 km/h.
  a Draw a graph of the journey.
  b When does the train arrive back at the station?

9 The graph below shows part of a cycle journey for a boy.

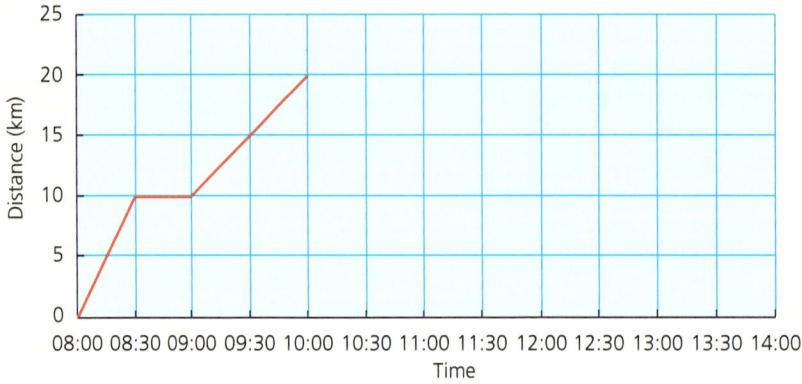

Copy the graph above.
At 10:00 the cyclist decides to cycle back to the start point.
The average speed of the cyclist travelling back to the start is 5 km/h.
He gets back in two phases, each with a different constant speed.
  a Draw two possible lines for the two phases of the return journey.
  b Calculate the speed of each of the two phases of the journey back.

Line graphs

So far, all the graphs have related to motion and the rate of change of distance with time. However graphs of other changes over time are also commonly used.

These are often plotted as a **line graph** with time along the x-axis.

Worked examples

1 The temperature (in °C) at a ski resort is recorded every 4 hours.
  The table shows the results for one of the days.

Time	00:00	04:00	08:00	12:00	16:00	20:00	24:00
Temperature (°C)	−8	−11	−6	3	4	1	−5

a   Display the data as a line graph.
    The points are joined by straight lines unless stated otherwise.

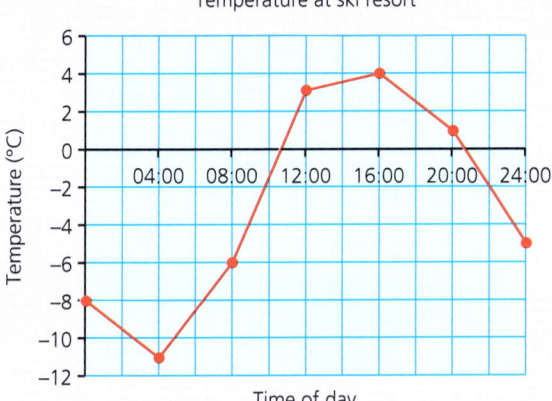

b   Use the graph to estimate the temperature in the resort at 10:00.

Although only seven temperature readings were taken, it is reasonable to assume that the temperature changed in a constant way between the readings. That is, because the temperature was −6 °C at 08:00 and 3 °C at 12:00, we assume that the temperature rose steadily between the two readings.

We draw a line vertically from 10:00 until it meets the graph, and then draw a horizontal line from this point to meet the temperature axis:

The temperature at 10:00 can therefore be estimated as approximately −1.5 °C.

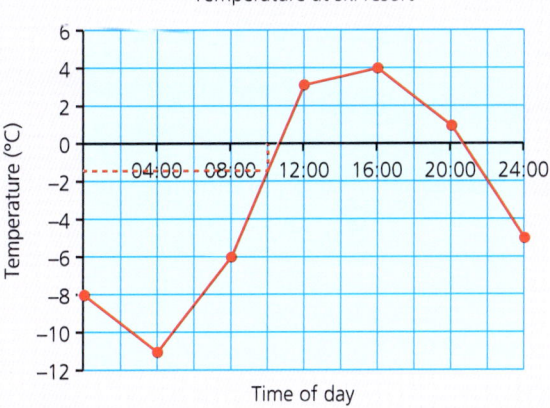

c   During which time is the rate of change of temperature increasing most rapidly? Justify your answer.

The rate of change is greatest, when the gradient (slope) of the graph is steepest. The graph is steepest between 08:00 and 12:00.

SECTION 3

2 The following two vessels are filled with water at the same constant rate.

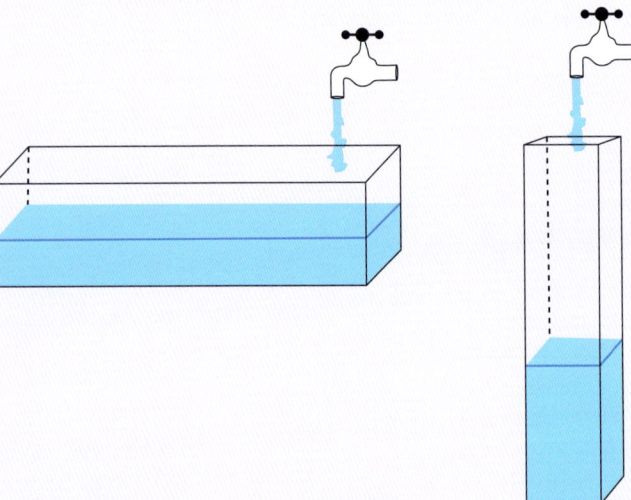

The depth of water is measured against time. Which of the graphs below is for which container? Justify your choice.

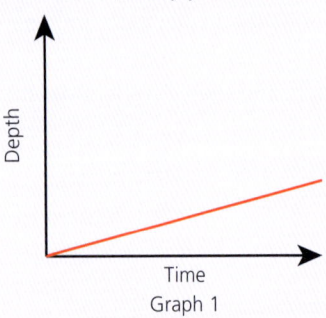

Graph 1

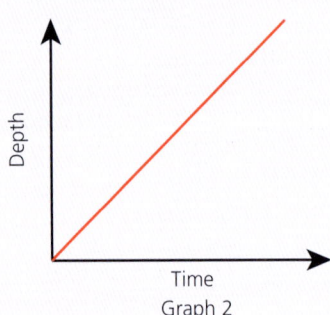

Graph 2

Graph 1 is for container A and Graph 2 for container B.

This is because for container A the rate of change of water depth with time will be slower than for container B, i.e. the depth will increase more slowly.

This is shown in the graph because the steepness of the line is less than that in Graph 2.

30 Graphs and rates of change

Exercise 30.2

1 A stadium is holding a festival. When the gates are opened, the total number of people in the stadium is recorded every 20 minutes over a 3-hour period. The results are shown in the table.

Time (mins)	0	20	40	60	80	100	120	140	160	180
Total number of people	0	3000	10 000	18 000	22 000	36 000	40 000	44 000	45 000	46 000

  a Plot a line graph to show the total number of people in the stadium over the 3-hour period.
  b Use your graph to estimate the number of people in the stadium after $1\frac{1}{2}$ hours.
  c Use your graph to estimate the number of people who entered the stadium between the 50th and 90th minutes after the gates opened.

2 In a café, the number of coffees sold is recorded every hour and the total number sold since the start of the day is calculated. The café opens at 08:00. The totals are shown in the table below.

Time	09:00	10:00	11:00	12:00	13:00	14:00	15:00	16:00
Total number of coffees sold	50	82	100	148	240	275	301	313

  a Plot a line graph of the data.
  b Use your graph to estimate the number of coffees sold by 11:15.
  c In which hour of the day are most coffees sold? Explain how you can tell this from your graph.
  d Use your graph to estimate the number of coffees sold between 11:30 and 14:30.

3 A swimming pool is left to heat up over a weekend. Unfortunately, the thermostat does not work properly and the pool overheats. The temperature of the water at 09:00 on Monday morning is 56 °C.
The heater is switched off and temperature readings are taken every 6 hours over the next three days. The results are shown below.

Day	Monday			Tuesday				Wednesday		
Time	09:00	15:00	21:00	03:00	09:00	15:00	21:00	03:00	09:00	15:00	21:00
Temperature (°C)	56	40	32	27	23	21	20	19.5	19	18.5	18

  a Plot a line graph of the data and draw a smooth curve through the points.
  b Estimate the temperature of the pool at 18:00 on Monday evening.

SECTION 3

c The pool can only be opened to the public once the temperature has dropped to 25 °C. Estimate the time at which the pool could have been opened.

d At what time was the temperature of the pool dropping the fastest? Explain, with reference to your graph, how you reached your answer.

4 The five containers below are filled with the same constant rate of water from a tap

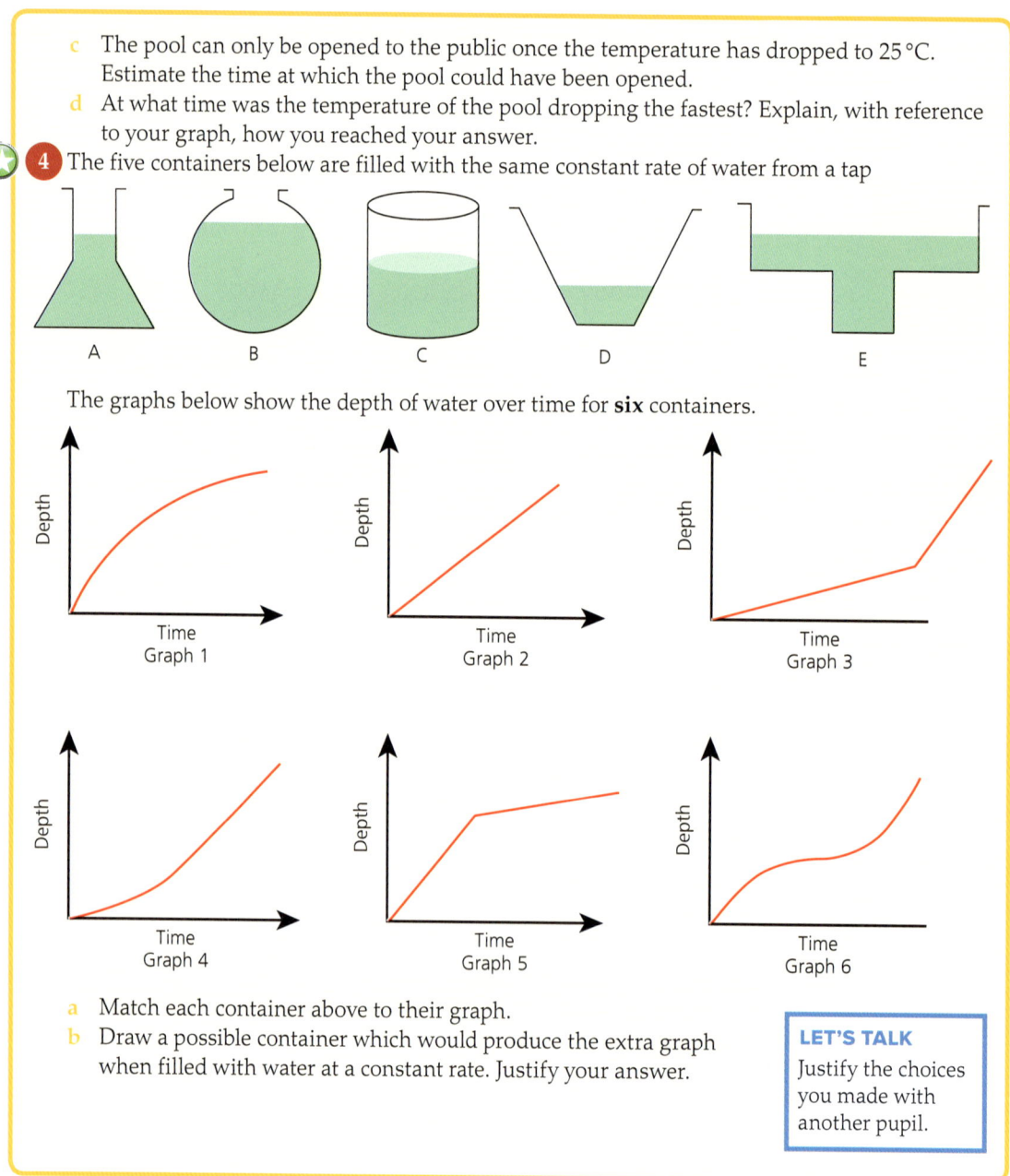

The graphs below show the depth of water over time for **six** containers.

a Match each container above to their graph.
b Draw a possible container which would produce the extra graph when filled with water at a constant rate. Justify your answer.

LET'S TALK
Justify the choices you made with another pupil.

▶ Now you have completed Unit 30, you may like to try the Unit 30 online knowledge test if you are using the Boost eBook.

Section 3 – Review

1. The first five numbers of a sequence are:
   $$-7 \quad -6 \quad -5 \quad -4 \quad -3$$
   a Write down the next two terms in the sequence.
   b Write down the nth term of the sequence.
   c What is the 100th term of the sequence?

2. In a year group of 120 students, students can opt to study either Spanish (S), Mandarin (M), neither language or both.
   The numbers of students studying each are shown in the Venn diagram below.

   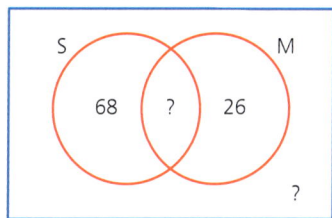

   a If 42 students study Mandarin, calculate the percentage of students who study Spanish.
   b What percentage of students do not study either language?

3. An octahedron is a 3D shape with eight faces each the shape of an equilateral triangle as shown:

   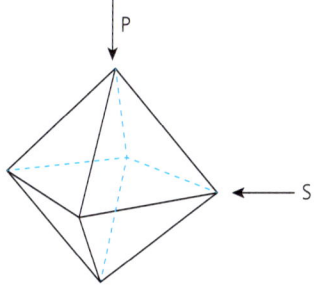

   a Sketch the 2D plan view indicated by arrow P.
   b Sketch the 2D side view indicated by arrow S.

4. A daughter is 27 years younger than her father. Let the daughter's age be d and the father's age F.
   Write an algebraic function machine to work out the daughter's age from the father's age.

SECTION 3

5 The coordinates of three vertices of a parallelogram ABCD are shown on the axes.

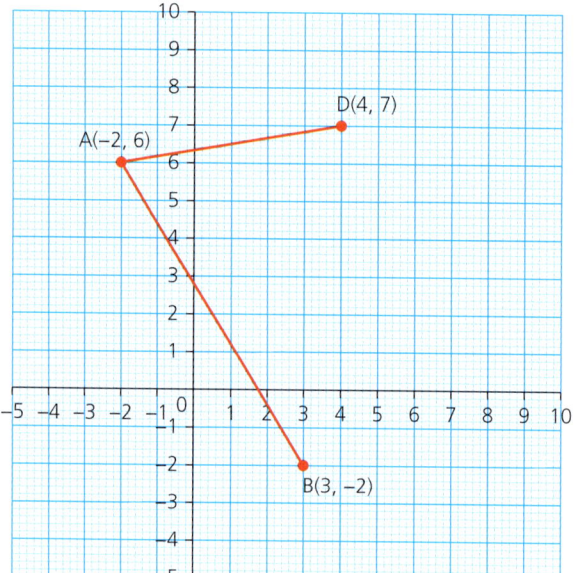

Calculate the coordinates of vertex C.

6 Forty-nine square tiles are arranged to form a square.
Show, by drawing three squares, that 49 can be written as the sum of three square numbers.

7 $y = x - 2$ and $y = 3x$ are the equations of two straight lines.
The coordinates of six points are given below:
A(4, 8)   B(3, 9)   C(−6, −4)   D(10, 30)
E(5, 3)   F(−1, −3)
Identify which line(s) the points lie on, by copying the Venn diagram below and writing the points in the correct region.

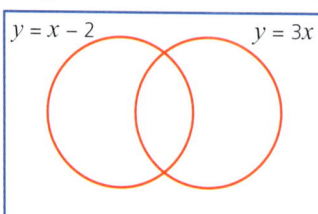

Section 3 – Review

8 A map is printed to a scale of 1 : 25 000.
   A rectangular field is 820 m × 1460 m in real life.
   Calculate its dimensions on the map, giving your answers in millimetres.

9 To make eight bread rolls, the following ingredients and amounts are needed:

   > 500g of bread flour   2 teaspoons of salt
   > 7g of yeast   300ml of warm water
   > $1\frac{1}{2}$ teaspoons of sugar   3 tablespoons of olive oil

   On checking the kitchen cupboard, Mario realises he has 1200 g of flour and 20 g of yeast.
   Assuming he has enough of the other ingredients, what is the maximum number of bread rolls he can make?

10 Two neighbours have identical water butts in their gardens to water their plants. One neighbour decides to water his plants by filling up a watering can several times. The other neighbour attaches a simple pump and hose to his water butt so that he can water his plants in one go.
   Both water butts are full to start with and the depth of water against time for both neighbours is shown on the two graphs X and Y below.

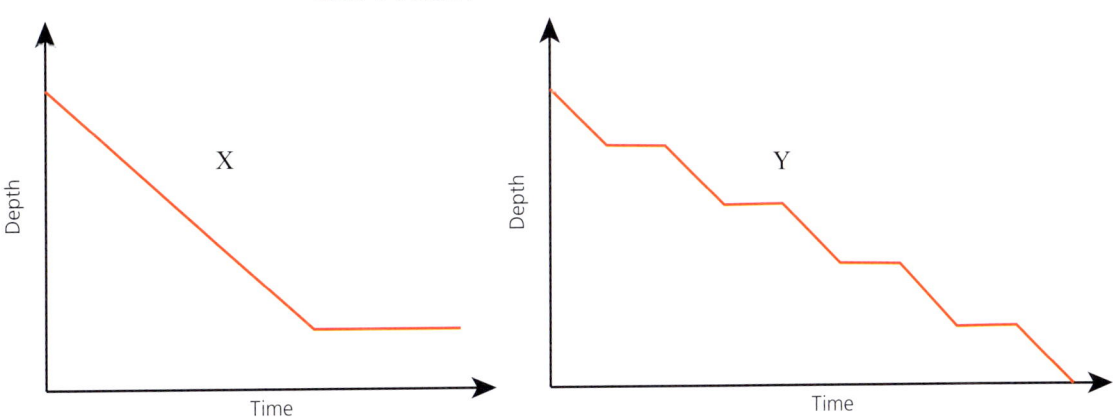

   a i) Which graph shows the neighbour using the watering can?
     ii) Justify your answer.
   b i) Which graph shows water left in the water butt after having watered the plants?
     ii) Justify your answer.

Glossary

Algebra Algebra is used in maths when we do not know the exact number(s) in a calculation. In algebra we use letters to represent unknown values or values that can change.

Arc Part of the circumference of a circle.

Bar graph A graphical representation of data using rectangular 'bars' where the height (or length) of the vertical (or horizontal) bars represent the frequency.

Carroll diagram A way of sorting objects, numbers and shapes by their traits. It looks like a table.

Categorical data Data that is used as labels rather than quantities, such as a favourite colour.

Centre of rotation A point about which an object with rotational symmetry is turned.

Chance How likely it is that some event will occur.

Chord A line segment joining one point of the circumference of a circle to another. If the chord also goes through the centre then it is a diameter.

Circumference The distance around the outside of a circle.

Coefficient The number with which we multiply the variable in question.

Commutative Multiplication is commutative, which means that changing the order of the calculation will not affect the answer, e.g. $3 \times 4 = 4 \times 3$.

Composite shape A shape that is made up of two or more common shapes.

Compound bar chart A graph that combines two or more types of information in one chart. It can also compare different quantities. It is a type of bar chart where columns can be split into sections to show breakdown of data.

Congruent Two shapes are congruent if they are exactly the same size and shape.

Constant A quantity whose value stays the same, as opposed to a variable whose value is variable.

Continuous data Data that can take any values. Examples include time, height and mass. Because continuous data can take any value, there are an infinite number of possible values.

Cube number A number that is made by multiplying an integer by itself three times. For example, $5 \times 5 \times 5 = 125$. Therefore 125 is a cube number.

Cube root To find the cube root of a number, work out which value multiplied by itself three times produces that number. For example, the cube root of 64 = 4 because $4 \times 4 \times 4 = 64$. This is written as $\sqrt[3]{64} = 4$.

Data Information collected from observations and measurements in statistics.

Decimal places The amount of numbers written after a decimal point.

Derive To obtain an answer by reasoning.

Diameter The straight line going through the centre of a circle, connecting two points on the circumference.

Direct proportion Two quantities are in direct proportion when they increase or decrease in the same ratio.

Glossary

Discrete data Numerical data that cannot be shown in decimals, for example, the number of children in a classroom.

Divisible A number that can be divided by another number exactly. For example, 24 is divisible by 6 because 24 ÷ 6 = 4.

Edge The line formed where two faces of a three-dimensional object meet.

Elevation Two-dimensional views of three-dimensional objects. For example, a front elevation is the view from the front.

Enlargement A type of transformation. All lengths of the original shape are multiplied by the same value known as the scale factor of enlargement.

Equal When two quantities or two mathematical expressions have the same value.

Equally likely outcome Events that have an equal chance of happening, such as getting a head or tail when flipping a coin.

Equation A mathematical statement stating that two quantities are the same.

Equivalent fractions Each fraction is worth the same amount.

Evaluate To calculate the value of.

Expand To expand an *expression* means to write it without brackets.

Experimental probability The chance of an event happening calculated by conducting an experiment.

Expression A mathematical statement written using algebra. It does not include an equals sign. For example, $3(x + y)$.

Face A single flat surface on a three-dimensional shape.

Factor A number that goes into another number exactly. For example, 5 is a factor of 20 because 20 ÷ 5 = 4.

Formula A rule that shows the relationship between different variables. For example, the formula for the area of a triangle is $A = \frac{1}{2}bh$, where b is the base length and h the perpendicular height of the triangle.

Frequency diagram A graph that shows the frequency (amount of times) of different groups.

Frequency table A table that lists items and shows the number of times the items occur.

Function A function relates an input to an output.

Function machine A diagram that represents a function. It shows an input, a mathematical operation and the output.

Grouped frequency table A method of organising large amounts of data into smaller groups.

Hectare A unit of area usually used for land. 1 hectare is equivalent to the area of a square of side length 100 m. So 1 hectare = 100 m × 100 m = 10000 m².

Highest common factor (HCF) The largest number that is a factor to two (or more) other numbers.

Image The name of a shape after its position has been transformed (changed) by rotation, reflection or enlargement.

Improper fractions A fraction where the numerator is bigger than the denominator.

Index (indices) The index of a number is the power it is raised to, for example, 2^4 where the index is the 4.

Inequalities A statement where two quantities are not necessarily equal to each other. For example, $x < 2$ means x has a value less than 2.

Infographic A type of graph or chart that uses pictures to make the information easier to understand.

Integer A whole number (not a fraction) that can be positive, negative or zero. Therefore, the numbers 10, 0, −25, and 5148 are all integers.

Glossary

Inverse The opposite of another operation. So the inverse of multiplication is division.

Likelihood The chance that something will happen.

Like terms Terms whose variables are the same. For example, $7x$ and $2x$ are like terms because they are both 'x'.

Line graph A graph using a line that shows information and how it changes (usually over time).

Long multiplication A type of written method for doing multiplications.

Lowest common multiple (LCM) The lowest common multiple of two numbers is the smallest whole number which both go into.

Mapping diagram A mapping diagram consists of two columns. One is for inputs and the other is for outputs. It shows how they are paired up.

Mean A type of average. Add all the values together and divide by how many numbers there are.

Median A type of average. It is calculated by writing all the numbers in order and then selecting the middle number.

Mixed number A fraction that consists of a whole number and a proper fraction.

Mode A type of average. It is the number that occurs the most often. Also known as the modal value.

Multiple A number that can be divided by another number a certain number of times without a remainder. For example, 24 is a multiple of 6 because $24 \div 6 = 4$.

Mutually exclusive Outcomes that cannot happen at the same time, for example getting a head and a tail when flipping a coin. You cannot get both at the same time.

***n*th term** See Position-to-term rule.

Number bonds The pairs of numbers that are added to make another number. For example, number bonds to 100 are pairs of numbers that add to 100, such as $35 + 65$.

Object An object is moved from its original position to a new position in a transformation.

Obtuse An obtuse angle is one between 90° and 180°.

Origin The point where the x- and y-axes cross. It has coordinates (0, 0).

Perpendicular Perpendicular lines are at right angles to each other.

Perpendicular height The distance measured at right angles from the base of a shape to its highest point.

Pie chart A type of graph in which a circle is divided into sectors that each represent a fraction of the whole.

Plan view A view from above.

Point of intersection The location where two lines cross.

Polygon A flat two-dimensional shape with straight sides. Quadrilaterals, triangles and hexagons are all examples of polygons.

Population The set of members under study where samples are taken.

Position-to-term rule A rule related to a term's position in a sequence. This is also known as the *n*th term rule.

Possible outcomes Results that could happen.

Power of 10 A power of 10 occurs when 10 is multiplied by itself a number of times.

Probability A quantitative measure of the likelihood of an event's occurrence, taking a value between 0 and 1 (inclusive).

Proof by counter example A way of showing that a given statement cannot be correct by showing an instance that contradicts it.

Glossary

Quantitative data Data representing an amount that can be measured, such as mass, height, number of people etc.

Questionnaires A list of questions that people have to answer. The questions often involve tick boxes.

Radius (radii) A straight line from the centre of a circle to its circumference.

Range A measure of how spread out the data is, i.e. the largest value – the smallest value.

Ratio How two or more quantities are related to each other.

Reciprocal The reciprocal of a number is 1 divided by that number. Therefore, the reciprocal of 3 is $\frac{1}{3}$.

Reflection symmetry A type of symmetry where one half is the reflection of the other half.

Regular polygon A polygon where all interior angles are the same and all edges are of the same length.

Relative frequency How often something happens divided by the total number of possible outcomes.

Remainder An amount left over after division. It happens when one number does not go into another exactly.

Representative sample A group chosen from the population with the same features as the population.

Rotational symmetry When an object is rotated around a centre point (turned) a number of degrees, the object appears the same. The order of symmetry is the number of positions the object looks the same in a 360° rotation.

Rounded Rounding a number means to write it in a simpler way, usually to make calculations easier to do.

Sample A group chosen from a population.

Sample size The size of the group chosen from a population.

Scale factor Used to describe an enlargement. The scale factor shows how much each length has been multiplied by.

Scatter graph A graph where two variables are plotted against each other and are represented by a point, for example a student's height plotted against their mass.

Sector An area of a circle enclosed by two radii and an arc.

Segment A section formed between an arc and a chord.

Sequence A set of numbers that follows a particular rule, such as 3, 6, 9, 12, 15 etc.

Simplest form Something written in its simplest from (such as a fraction or ratio) is when each part cannot be reduced further and still be equivalent.

Solve To determine the solutions to a problem.

Square number The result of multiplying a whole number by itself. For example, $5 \times 5 = 25$ so 25 is a square number.

Square root The square root of a number is to work out which value multiplied by itself produces that number. For example, the square root of 64 = 8 because $8 \times 8 = 64$. This is written as $\sqrt{64} = 8$.

Squaring Multiplying a number by itself.

Tally A small mark used to record that an outcome has happened. It is usually displayed in a table.

Tangent A straight line that touches the circle at only one point.

Terminating decimals A decimal that ends after a certain number of decimal places, e.g. 0.2, 3.625.

Glossary

Term-to-term rule A rule that describes how to get from one term to the next in sequence.

Terms Numbers in a **sequence**.

Tesselate Shapes that tessellate fit together without leaving any gaps.

Theoretical probability The likelihood of something happening in theory. For example, the theoretical probability of getting a tail when flipping a coin is $\frac{1}{2}$.

Transformation When an object undergoes a transformation its size and position may change, for example through reflection, rotation, enlargement or translation.

Translation Translation is a sliding movement. To describe a translation, you need to give how far the shape has moved horizontally and how far it has moved vertically.

Truncated cone A cone with the top sliced off.

Two-way table A statistical table that shows the number or frequency for two **variables**, the rows indicating one category and the columns indicating the other category.

Unbiased An event is unbiased if it is fair.

Variables A mathematical quantity whose value can vary.

Venn diagram A diagram that uses circles to represent sets, in which the relations between the sets are indicated by the arrangement of the circles.

Vertex (vertices) A corner of a two-dimensional or three-dimensional shape. The plural of vertex is vertices.

Waffle diagram Similar in use to a **pie chart** but it uses a grid format to represent percentages rather than parts of a circle.

x**-axis** The horizontal axis.

y**-axis** The vertical axis.

Index

A
addition
  borrowed numbers 3
  fractions 110, 119–20
  mixed numbers 121
  number bonds 2
  order of operations 25, 26, 31
algebra
  equations 163–5
  expressions 28–32, 33, 109–15, 153–5
  formulae 32, 33–5, 74, 77, 155–7
  functions 247
angles 135–52, 169
  around a point 145–8
  drawing and measuring 135–8
  geometric figures 139–42
  parallel lines 148–52
  quadrilaterals 138–9, 142–4
area
  compound shapes 22–4
  cuboid surface area 60–3
  decimals 128
  expressions 112, 114
  formulae 74
  triangles 19–22
  units of 225–6
averages 87–91, 158
axes 100–2, 198, 201, 242

B
bar graphs 42–6
Bernoulli, Jakob 158
BIDMAS 25, 31
borrowed numbers 3
brackets 25, 26, 27, 31, 112–13

C
calculator, using a 78, 192, 206, 209
capacity, metric units 222, 223–5
Carroll diagrams 40–1, 42
categorical data 14, 36
centre of rotation 95, 103–5
circles 11, 12, 97
coefficients 28
common denominator 119–20
composite shapes 58–60, 114
compound bar charts 43–4
compound shapes 22–4
cones 191
congruent shapes 148
continuous data 13, 14, 36, 37–8, 44
conversion of units 223–5
coordinates 198–204, 211–20, 248
cube numbers 209
cube roots 209–10
cubes 54–5, 56, 75, 186–9, 208–9
cuboids 56–8, 60–3
cylinders 54–5, 190

D
data
  collecting 13–18
  organising 36–9
  presenting 40–53
decimal places 13, 78–80
decimals 80–6, 116–19
degrees 135
denominator 110, 116–17, 119–20, 122, 124
direct proportion 234–6
discrete data 13, 36, 37, 44
distance-time graphs 237–42
divisibility 65–9, 76, 124
division 5–6
  cube roots 209–10
  decimals 83, 85
  fractions 125–6
  negative numbers 6–7
  order of operations 25, 31
  powers of 10 82
  square roots 206–8

E
edges 54–5
elevations 185–91
enlargement 106–8, 227
equation of the straight line 211, 213, 217–19, 248
equations 28–9, 32, 162–7
equilateral triangles 9, 10, 22, 95, 96, 112
equivalent ratios 230–3
estimation 3–5, 80–6
experimental probability 158–61, 169
expressions 28–32, 33, 109–15, 153–5

F
faces 54–5
factors 64–5
formulae 32, 33–5, 74, 77, 155–7
fractions 110, 116–29, 169, 179–80, 230
frequency diagrams 42–6
frequency tables 36–9, 44, 91–4
functions 192–7, 211–20, 247

G
geometric figures, constructing 138–42
geometry 171
graphs
  line 48–9, 242–6, 249
  linear functions 211–20
  scatter 49–50, 53
  travel 237–42
grid method 84–5, 86
grouped frequency tables 37, 44

H
hexagons 8, 11, 95, 97
highest common factor (HCF) 64–5
history of mathematics 1, 77, 171

I
improper fractions 121, 122
indice1.5s 25, 31
inequalities 166–7
infographics 51–2
integers 5, 205, 209
interviews 16, 18, 73
inverse function 193, 194–5
isosceles trapeziums 96, 144
isosceles triangles 9, 97

Index

K
kites 8–9, 96, 144, 170

L
law of averages 158
length, metric units 221, 223, 224, 228–9
line graphs 48–9, 242–6, 249
linear functions 211–20
long multiplication 83, 86
lowest common multiple (LCM) 64–5, 119, 122

M
mass, metric units 222, 223, 224
mean 87–8, 89, 90–4, 168
measurement, units of 221–9
median 88, 89, 90–4, 168
mental skills 2–3
metric system 221–6
mixed numbers 121–3
mode 88, 89, 90–4
multiples 64–5
multiplication 5–6
   cube numbers 209
   decimals 83–6
   fractions 123–4, 125–6, 127
   negative numbers 6–7
   order of operations 25, 26, 31
   powers of 10 81–2
   simple expressions 112
   square numbers 205–6

N
negative numbers 3, 6–7
nth term 176–7
number bonds 2
numerator 116–17, 122, 124

O
obtuse angles 142
octagons 8, 10
order of operations 25–7, 31
origin 198
outcomes 70–1, 130–4

P
parallel lines 148–52
parallelograms 8–9, 96, 138, 144, 148, 152
pentagons 8, 97
percentages 116–18, 179–84
perimeter 28, 30–1, 33–5, 111, 114, 168
pie charts 46–8, 75, 233, 236
place value 1, 116
plan view 185–6, 190–1, 247
point, angles around a 145–8
polygons 8–11
populations 14, 15
position-to-term rule 176–7
prisms 59
probability 70–2, 76, 130–4, 158–61, 169
proportion 234–6
pyramids 54–6, 190

Q
quadrilaterals 138–9, 142–4
quantities 179–84
questionnaires 16–18, 73

R
range 89–90, 93, 168
ratios 226, 230–6
reciprocals 125
rectangles 8–9, 33–4, 128, 168
   angles 139–40
   expressions 28–31, 111–15
   symmetry 96
   translation 204
reflective symmetry 95–102
relative frequency 158
remainder 85
rhombuses 8–9, 73, 96, 199–200
right angles 19, 20, 139
rotation 103–5
rotational symmetry 9–11, 73, 95–9, 168
rounding 3, 4, 78–9

S
sampling 14, 15
scale drawings 226–9
scale factor of enlargement 106–8, 227
scatter graphs 49–50, 53
sequences 172–8, 247
shapes
   algebra 29–31, 111
   composite 58–60, 114
   compound 22–4
   coordinates 198–204
   three-dimensional 54–63, 185–91
   two-dimensional 8–13, 95–108, 198–204
simplifying 31–2, 109–12, 126–7, 231
square numbers 205–6, 248
square roots 206–8
squares 8–9, 10, 205
   angles 139–40
   area 24
   symmetry 96
   translation 204
substitution 32–4
subtraction
   borrowed numbers 3
   fractions 119–20
   negative numbers 3
   number bonds 2
   order of operations 25, 26, 31
surface area 60–3
symmetry 9–11, 73, 95–9, 168

T
tables 36–9, 44, 91–4
tally and frequency tables 36–9
term-to-term rules 174–5
tessellation 148, 152
time 237–46, 249
transformations 106
translation 106, 202–4
trapeziums 8–9, 74, 96, 144
travel graphs 237–42
triangles 9–11, 74, 112
   area 19–24
   reflection 100, 102
   symmetry 95, 96, 97
   translation 203
two-way tables 38

U
units of measurement 221–9

V
Venn diagrams 40, 41, 131, 247, 248
vertices 11, 54–5, 199–204, 248
volume 56–8, 59–60, 75

W
waffle diagrams 46–8, 75